Pattern Mining with Evolutionary Algorithms

Sebastián Ventura • José María Luna

Pattern Mining with Evolutionary Algorithms

Springer

Sebastián Ventura
Department of Computer Science
  and Numerical Analysis
University of Cordoba
Cordoba, Spain

José María Luna
Department of Computer Science
  and Numerical Analysis
University of Cordoba
Cordoba, Spain

ISBN 978-3-319-81618-0   ISBN 978-3-319-33858-3   (eBook)
DOI 10.1007/978-3-319-33858-3

Printed on acid-free paper

This Springer imprint is published by Springer Nature
The registered company is Springer International Publishing AG Switzerland

–Success is a journey, not a destination. The doing is often more important than the outcome–

Arthur Ashe.

To our families.

Preface

This book is intended to provide a general and comprehensible overview of the field of pattern mining with evolutionary algorithms. To do so, the book provides formal definitions about patterns, pattern mining, type of patterns, and the usefulness of patterns in the knowledge discovery process. As it is described within the book, the discovery process suffers from both high runtime and memory requirements, especially when high-dimensional datasets are analyzed. To solve this issue, many pruning strategies have been developed. Nevertheless, with the growing interest in the storage of information, more and more datasets comprise such a dimensionality that the discovery of interesting patterns becomes a hard process. In this regard, the use of evolutionary algorithms for mining pattern enables the computation capacity to be reduced, providing sufficiently good solutions.

The book also provides a survey on evolutionary computation with particular emphasis on genetic algorithms and genetic programming. Additionally, this book carries out an analysis of the set of quality measures most widely used in the field of pattern mining with evolutionary algorithms. This book serves as a good review on the most important evolutionary algorithms for pattern mining. In this sense, it considers the analysis of different algorithms for mining different types of patterns and relationships between patterns, such as frequent patterns, infrequent patterns, patterns defined in a continuous domain, or even positive and negative patterns.

The book also introduces a completely new problem in the pattern mining field, which is known by the name of the mining of exceptional relationships between patterns. In this problem, the goal is to identify patterns where distribution is exceptionally different from the distribution in the complete set of data records. Finally, this book deals with the subgroup discovery task, a method to identify a subgroup of interesting patterns that is related to a dependent variable or target attribute. This subgroup of patterns satisfies two essential conditions: interpretability and interestingness.

Cordoba, Spain
February 2016

Sebastián Ventura
José Mará Luna

Acknowledgments

We would like to thank the Springer editorial team for giving us the opportunity to publish this book, for their great support toward the preparation and completion of this work, and for their valuable editing suggestions to improve the organization and readability of the manuscript. We also want to thank our colleagues for their valuable help during the preparation of the book, whose comments were very helpful for improving its quality.

This work was supported by the Spanish Ministry of Economy and Competitiveness under the project TIN2014-55252-P and FEDER funds.

Contents

Chapter 1
Introduction to Pattern Mining

Abstract This chapter introduces the pattern mining task to the reader, providing formal definitions about patterns, the pattern mining task and the usefulness of patterns in the knowledge discovery process. The utility of the extraction of patterns is introduced by a sample dataset for the market basket analysis. Different type of patterns can be considered from the pattern mining point of view, so an exhaustive taxonomy about patterns in this field is presented, describing concepts such as frequent and infrequent patterns, positive and negative patterns, patterns expressed in compressed forms, sequential patterns, spatio-temporal patterns, etc. Additionally, some pruning strategies to reduce the computational complexity are described, as well as some efficient pattern mining algorithms. Finally, this chapter formally describes how interesting patterns can be associated to analyse the causality by means of association rules.

1.1 Definitions

The exponential increasing amounts of data generated and stored in many different areas brings about the need for analysing and getting useful information from that data. Generally, raw data lacks of interest and an in-depth analysis is required to extract the useful knowledge that is usually hidden. This process has giving rise to the field known as knowledge discovery in databases (KDD) [12].

KDD is concerned with the development of methods and techniques for making sense of data, and the discovery of patterns plays an important role here. Under the term *pattern* we can define subsequences, substructures or itemsets that represent any type of homogeneity and regularity in data [1]. Thus, patterns represent intrinsic and important properties of datasets.

Given a set of items $I = \{i_1, i_2, \ldots, i_n\}$ in a dataset, a pattern P is formally defined as a subset of I, i.e. $\{P = \{i_j, \ldots, i_k\} \subseteq I, 1 \leq j, k \leq n\}$, that describes valuable features of data. Given a pattern P, the length or size of that pattern is expressed as $|P|$, denoting the number of single items or singletons that it includes. Thus, the length of a simple pattern $P = \{i_1, \ldots, i_j\} \subseteq I$ is defined as $|P| = j$ since it is comprised of j singletons.

© Springer International Publishing Switzerland 2016
S. Ventura, J.M. Luna, *Pattern Mining with Evolutionary Algorithms*,
DOI 10.1007/978-3-319-33858-3_1

Fig. 1.1 Analysis of the market basket, including five different customers

The task of finding and analysing patterns in a database might be considered as straightforward [2], but it gets increasingly complicated once the interest of the discovered patterns becomes a priority. Pattern mining is considered as an essential part of the KDD process, being defined as the task of discovering patterns of high interest in data for a specific user aim. Thus, the process of quantifying in a proper way the interest of the discovered patterns includes different metrics that are highly related to the purpose for which the task is applied [9]. Market basket analysis is perhaps the first application domain in which mining patterns of interest was useful. In many occasions, a high number of purchases are bought on impulse [6] and it is required an in-depth analysis to obtain clues that let us know which specific items are strongly related. For instance, how likely is for a customer to buy bananas and milk together? In this regard, it seems easy to determine that the interest of the patterns might be quantified by their probability of occurrence [12]. Thus, considering a sample set of customers (see Fig. 1.1), we can obtain that the pattern $P = \{Banana, Milk\}$ has a frequency of three over a total of five customers, so there exists a probability of 60 % that a random customer buys both bananas and milk in the same trip. This analysis could allow shopkeepers to increase the sales by re-locating the products on the shelf, or even it could allow managers to plan advertising strategies.

Pattern mining results can be analysed in many different ways, and the same pattern might be used to follow different marketing strategies. For instance, the aforementioned pattern $P = \{Banana, Milk\}$ might be used to understand that placing these products in proximity produces an encouragement of the sales of both products at time. Any customer that buys any of these items will have the temptation of buying the other one. Nevertheless, the same pattern P might be analysed differently, and both products might be placed far from each other. In this regard, since both products tend to be purchased together, the fact of a customer going over the supermarket may give rise to buy something else on impulse.

According to some authors [1], pattern mining is a high level and challenging task due to the computational challenge issues. The pattern mining problem turns into an arduous task depending on the search space, which exponentially grows with the number of single items or singletons considered into the application domain. For a better understanding, let us consider a set of items $I = \{i_1, i_2, \ldots, i_n\}$ in a dataset. A maximum number of $2^{|I|} - 1$ different patterns can be found, so a straightforward approach becomes extremely complex with the increasing number of singletons. As a matter of example, let us consider that, in average, about 8,000 grains of sand fit in a cubic centimetre. Hence, it is estimated that there are about 10×10^{36} grains of sand on all the world's beaches. This value is lower than the number of patterns that can be found in a dataset comprising 150 singletons, i.e. $2^{150} - 1 \approx 1.42 \times 10^{45} \gg 10 \times 10^{36}$.

1.2 Type of Patterns

Since its introduction [5], the pattern mining task has provoked tremendous curiosity among experts on different application fields [26, 28]. This growing interest encouraged the definition of new type of patterns according to the analysis required by experts on different application field. The following discussion will address and describe each of these patterns.

1.2.1 Frequent and Infrequent Patterns

Market basket analysis is considered as the first application domain in which the pattern mining task plays an important role. In this domain, the mining of strongly related products is essential to take right decisions to increase the sales, so the mining of frequent products purchased together is the key [3]. To quantify this frequency, it is calculated as the number of transactions in which the set of products is included in a dataset. Table 1.1 illustrates an example dataset based on five different customers (see Fig. 1.1). Each row corresponds to a different transaction that represents a customer, and each transaction contains a set of items that describes the products purchased by the customer.

Let $I = \{i_1, i_2, \ldots, i_n\}$ be a set of items in a dataset, and let $T = \{t_1, t_2, \ldots, t_m\}$ be the set of all dataset transactions. Each transaction t_j is a set of items such that $t_j \subseteq I$. Let P also be a pattern comprising a set of items, i.e. $P \subseteq I$. The pattern P satisfies the transaction t_j if and only if $P \subseteq t_j$, and the frequency of this pattern $f(P)$ is defined as the number of different transactions that it satisfies, i.e.

Table 1.1 An example of market basket dataset obtained from Fig. 1.1

Customer	Items
ID_1	{Banana, Bread, Butter, Orange}
ID_2	{Banana, Bread, Milk}
ID_3	{Banana, Butter, Milk}
ID_4	{Bread, Butter, Milk, Orange}
ID_5	{Banana, Butter, Milk, Orange}

$f(P) = |\{\forall t_j \in T : P \subseteq t_j\}|$. A pattern P is defined as frequent if and only if the number of transactions that it satisfies is greater or equal to a minimum predefined threshold f_{min}, i.e. $f(P) \geq f_{min}$.

As previously analysed, a dataset comprising n singletons contains $2^n - 1$ different itemsets, whereas the number of itemsets of size k is equal to $\binom{n}{k}$ for any $k \leq n$. Thus, given the amount of computations needed for each candidate is $O(k)$, the overall complexity of the mining is $O(\sum_{k=1}^{n} k \times \binom{n}{k}) = O(2^{n-1} \times n)$. All of this led us to the conclusion that the complexity of finding patterns of interest is in exponential order, and this complexity is even higher when the frequency of each itemset is calculated [14]. For any dataset comprising n singletons and m different transactions, the complexity to compute the frequencies of all the patterns within the dataset is equal to $O(2^{n-1} \times m \times n)$.

Going back to the aforementioned example, a space of exploration of size $2^5 - 1 = 31$ can be arranged as a lattice (see Fig. 1.2) where the number of transactions satisfied by each itemset is illustrated into brackets. The number of k-itemsets (itemsets of size k) that can be found in a dataset that comprises n singletons is $\binom{n}{k} = \frac{n!}{k!(n-k)!}$. Thus, considering the sample market basket dataset, the number of 1-itemsets is $\binom{5}{1} = 5$; the number of 2-itemsets is $\binom{5}{2} = 10$; the number of 3-itemsets is $\binom{5}{3} = 10$; etc, obtaining that $\binom{5}{1} + \binom{5}{2} + \ldots + \binom{5}{5} = 2^5 - 1 = 31$, which is the whole space of exploration for this example.

Continuing with the same example, and considering a $f_{min} = 2$, the lattice is separated (dashed line) into two main parts: frequent and infrequent. Any pattern P that appears above the border that divides the search space into two parts is considered as a frequent pattern since $f(P) \geq f_{min} = 2$. Thus, according to this measure, which determines whether a pattern is of interest or not, the space of exploration can be reduced significantly. For instance, from a total of $2^5 - 1 = 31$ itemsets, the search space is reduced to a total of 19 frequent patterns when $f_{min} = 2$ is considered. The increasing of f_{min} implies a reduction of the search space.

As mentioned above, the process of mining frequent patterns obtains a set of patterns whose frequency overcomes the minimum threshold f_{min}. According to f_{min}, any frequent pattern represents the most valuable information that reflect intrinsic and important properties of datasets. However, there are situations where it is interesting to discover abnormal or unusual behaviour in datasets, discovering rare or infrequent patterns [15], i.e. those that do not follow the trend of the others [33]. A pattern P is defined as infrequent if and only if the number of transactions that

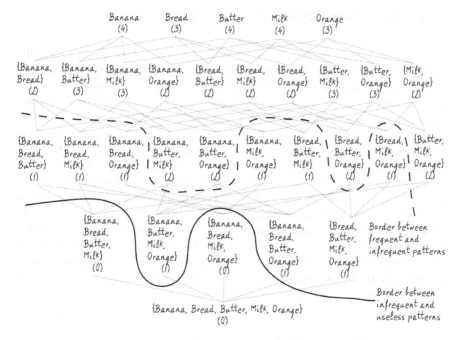

Fig. 1.2 Lattice analysis of the market basket, including five different customers

it satisfies is lower than a maximum predefined threshold f_{max}, i.e. $f(P) = |\{\forall t_j \in T : P \subseteq t_j\}| < f_{max}$. Nevertheless, this maximum threshold value implies that any itemset that does not appear in the dataset will also be considered as infrequent. This can be solved by including a minimum threshold that divides the set of non-frequent patterns into infrequent and useless patterns—those that can be discarded. Thus, a second definition of infrequent patterns can be described. A pattern P is defined as infrequent if and only if the number of transactions that it satisfies is lower than a maximum predefined threshold f_{max} and greater than a minimum threshold f_{min}, i.e. $f_{min} < f(P) < f_{max}$.

Again analysing the lattice (see Fig. 1.2) that comprises the itemsets of the sample market basket dataset illustrated in Table 1.1, and considering $f_{max} = 2$, and $f_{min} = 0$, it is obtained that the set of infrequent patterns comprises those patterns located between the dashed (border between frequent and infrequent patterns) and the solid (border between infrequent and useless patterns) lines. Thus, according to both thresholds, there are nine infrequent patterns in the dataset. Finally, we obtain three patterns that are not included in the dataset and they can be considered as useless.

1.2.2   Closed and Maximal Frequent Patterns

Many real-world application domains [22] contain patterns whose length is typically too high. The extraction of such lengthy patterns requires a high computational time when the aim is the extraction of frequent patterns. Noted that a major feature of any frequent pattern is that any of its subsets is also frequent [1], so a significant amount of time can be spent on counting redundant patterns. For instance, given a pattern P of length $|P| = 5$, it comprises a total of $2^5 - 2 = 30$ sub-patterns that are also frequent. In this regard, the discovery of condensed representations of frequent patterns is a solution to overcome the computational and storage problems.

Suppose a dataset whose set of frequent patterns is defined as $\mathscr{P} = \{P_1, P_2, \ldots, P_n\}$. A frequent pattern $P_i \in \mathscr{P}$ is defined as maximal frequent pattern if and only if it has no frequent superset, i.e. $\{P_i : \nexists P_j \supset P_i, P_j \in \mathscr{P} \land P_i \in \mathscr{P}\}$. The number of maximal frequent patterns is considerably smaller than the number of all frequent patterns. For the sample market basket dataset shown in Table 1.1 the set of maximal frequent patterns is illustrated in Fig. 1.3. As illustrated, there is no super-patterns for the pattern $P = \{Banana, Bread\}$ that satisfies the minimum frequency threshold $f_{min} = 2$, so P is defined as a maximal frequent pattern.

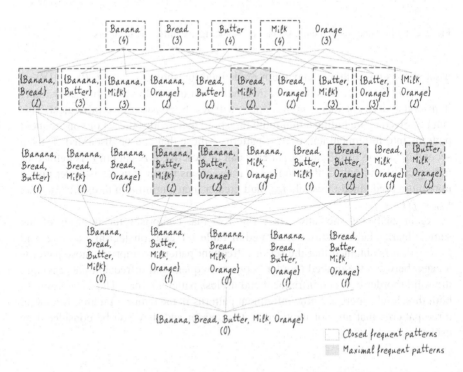

Fig. 1.3 Maximal and closed frequent patterns represented in the lattice of the market basket dataset

Fig. 1.4 Set of frequent,
closed and maximal patterns
in a dataset

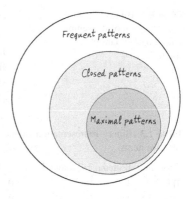

Condensed representations of frequent patterns can also be carried out by means
of closed frequent patterns. Continuing with the same set of frequent patterns
defined as $\mathscr{P} = \{P_1, P_2, \ldots, P_n\}$, a frequent pattern $P_i \in \mathscr{P}$ is defined as closed
frequent pattern if and only if it has no frequent superset with the same frequency
of occurrence that it has, i.e. $\{\nexists P_j \supset P_i : f(P_i) \geq f_{min} \wedge f(P_j) = f(P_i)\}$. Analysing
Fig. 1.3, the set of closed frequent patterns is illustrated by a dashed rectangle.
It should be noted that, since maximal frequent patterns have no frequent superset,
hence they are also closed frequent patterns (see Fig. 1.4). Nevertheless there is
a huge difference between them, since maximal patterns lose information. Exact
frequencies of the subsets cannot be directly derived from the frequency of the
maximal frequent patterns. On the contrary, closed frequent patterns do not lose
information, and the frequencies of the subsets can be derived from the set of closed
frequent patterns.

1.2.3 Positive and Negative Patterns

Heretofore, we have addressed the pattern mining task as the discovery of patterns
whose co-occurrence provides intrinsic and important properties of datasets. The
problem of mining this type of patterns is highly related to the metric used to
quantify the interest of the discovered patterns [35]. For instance, an item that rarely
appears in the database will give rise to itemsets with a low frequency, so items that
tend to be infrequent will form infrequent or useless patterns [34]. In this regard, it
is interesting to mine patterns in which presence as well as absence of an item may
be used.

Considering the sample market basket dataset described in Table 1.1, it can be
represented in a binary format as shown in Table 1.2, where each row represents a
particular transaction and each column describes a specific item. This data format
illustrates the representation of positive items, so a value one is obtained if the item
is present in a transaction and zero otherwise. Thus, in mining positive patterns the
presence of an item in a transaction is more important than its absence.

Table 1.2 Binary representation of the sample market basket dataset (Table 1.1)

Customer	Banana	Bread	Butter	Milk	Orange
ID_1	1	1	1	0	1
ID_2	1	1	0	1	0
ID_3	1	0	1	1	0
ID_4	0	1	1	1	1
ID_5	1	0	1	1	1

Table 1.3 Binary representation of the sample market basket dataset (Table 1.1) considering negative items

Customer	Banana	¬Banana	Bread	¬Bread	Butter	¬Butter	Milk	¬Milk	Orange	¬Orange
ID_1	1	0	1	0	1	0	0	1	1	0
ID_2	1	0	1	0	0	1	1	0	0	1
ID_3	1	0	0	1	1	0	1	0	0	1
ID_4	0	1	1	0	1	0	1	0	1	0
ID_5	1	0	0	1	1	0	1	0	1	0

In order to enhance the importance of infrequent items, which cannot be mined if a low frequency threshold f_{min} is considered, numerous methods have been proposed in the literature for finding negative relationships between items [16]. For instance, considering the frequency of a pattern $f(P)$ as the metric to determine the interest of such a pattern P, it is easy to determine that the pattern $P_i = \{Banana, Bread\}$ can be considered as frequent since it is satisfied by two transactions, i.e. $f(P_i) \geq f_{min} = 2$. This pattern P_i is described as a positive relationship between the purchase of both *Banana* and *Bread*. Nevertheless, negative relationships can also be described if the items are represented in a negative way (see Table 1.3), where symbol ¬ indicates negation. Thus, a negative pattern $P_j = \{Banana, \neg Bread\}$ describes a negative relationship between the purchase of both *Banana* and *Bread*, whose frequency $f(P_j) = 2$. It describes that two over five customers purchase *Banana* and do not purchase *Bread*. Noted that the frequency of a singleton represented in its negative form is equal to the number of transactions minus the frequency of the same item represented in its positive form, i.e. $f(\neg Banana) = 5 - f(Banana) = 1$.

The discovery of negative frequent patterns can be carry out similarly to the mining of positive frequent patterns [38]. Negative items are added to the dataset (see Table 1.3) and the mining process follows the same process as if all the singletons were positive. However, it should be noted that the use of negative items in many application domains implies the presence of items with an extremely high probability, which could provide that the extracted knowledge is meaningless. In such a situation, it is more recommendable to mine infrequent patterns than adding negative items to the database to mine negative relationships. As a matter of example, let us consider a medical database [24] including extremely unusual diseases. Here, the mining of infrequent patterns provides much more useful information than the one provided by mining frequent negative patterns—most of them will include the negation of rare diseases.

1.2.4 Continuous Patterns

Conceptually, the pattern mining problem can be viewed as the process of discovering relationships between items in a binary database (see Table 1.1) where the value "1" represents the presence of a specific item in a specific transaction, "0" otherwise. Nevertheless, many business domains gather data including items defined in continuous domains, which include richer information, e.g. age or salary. This type of items cannot be represented in a binary way, and it requires new techniques for discovering such patterns, giving rise to the concept of continuous patterns [38]. Unlike traditional patterns, continuous patterns can be represented as a range of values, including upper and lower limits.

Suppose a dataset comprising a set of items $I = \{i_1, i_2, \ldots, i_n\}$, where at least one item $i_j \in I$ is defined in a continuous domain $\mathcal{R}$, i.e. $i_j \in \mathcal{R}$. This type of items are called numerical items, since they represent values in a continuous domain. A pattern P is defined as a continuous pattern if it comprises at least one numerical item, i.e. $\{P \subseteq I : \exists i_j \in P \wedge i_j \in \mathcal{R}\}$. In pattern mining, a numerical item is represented as range of values $i_j = [x_l, x_u]$, x_l and x_u describing the lower and upper limits, respectively. As a matter of example, let us consider the aforementioned sample market basket dataset in which we have added the numerical item *Invoice* (see Table 1.4). A continuous pattern P over this dataset can be defined as $P = \{Banana, Invoice = [5.3, 5.9]\}$. Let $T = \{t_1, t_2, \ldots, t_m\}$ be the set of all dataset transactions, and each transaction t_j be a set of items such that $t_j \subseteq I$. A continuous pattern P satisfies the transaction t_j if and only if $P \subseteq t_j$, and the frequency of this pattern $f(P)$ is defined as the number of different transactions that it satisfies, i.e. $f(P) = |\{t_j : P \subseteq t_j, t_j \in T\}|$. Thus, analysing all transactions in data (see Table 1.4), the subset of transactions satisfied by the pattern $P = \{Banana, Invoice = [5.3, 5.9]\}$ comprises the transactions $\{t_2, t_3\}$.

The mining of continuous patterns is a challenging problem [11] due to the number of different numerical items is infinity according to the lower and upper limits. To solve this issue, many authors have proposed to transform the continuous and infinite search space into a discrete one by splitting the search space into intervals [32]. Thus, for the numerical item *Invoice*, it is possible to divide its range of values into a fixed number of intervals which minimum and maximum limits are 5.4 and 7.4, respectively. For instance, giving three different intervals,

Table 1.4 Sample market basket dataset that includes a continuous pattern

Customer	Banana	Bread	Butter	Milk	Orange	Invoice
ID$_1$	1	1	1	0	1	6.5
ID$_2$	1	1	0	1	0	5.4
ID$_3$	1	0	1	1	0	5.8
ID$_4$	0	1	1	1	1	6.6
ID$_5$	1	0	1	1	1	7.4

Table 1.5 Sample market basket dataset that includes a continuous pattern

Customer	Banana	Bread	Butter	Milk	Orange	Invoice [5.4, 6.0]	[6.1, 6.7]	[6.8, 7.4]
ID$_1$	1	1	1	0	1	0	1	0
ID$_2$	1	1	0	1	0	1	0	0
ID$_3$	1	0	1	1	0	1	0	0
ID$_4$	0	1	1	1	1	0	1	0
ID$_5$	1	0	1	1	1	0	0	1

the following "new" items can be obtained: [5.4, 6.0], [6.1, 6.7] and [6.8, 7.4]. All these new items can replace the numerical item *Invoice*, obtaining a new dataset (see Table 1.5) where traditional patterns can be discovered.

It should be noted that the resulting dataset is highly dependent on the defined intervals. Thus, the knowledge extracted by means of mining patterns of interest varies when continuous patterns are transformed into binary patterns or discrete patterns. As a matter of example, and considering the dataset illustrated in Table 1.5, the pattern $P = \{Banana, Invoice = [5.4, 6.0]\}$ satisfies two transactions, whereas the same two transactions can be also represented by a different pattern $P' = \{Banana, Invoice = [5.4, 6.3]\}$, $P \neq P'$. Thus, the same transactions may describe different information.

1.2.5 Colossal Patterns

The importance of mining patterns in datasets is entailing an increasing attraction to the research community, and many application domains are applying the extraction of patterns of interest to discover hidden knowledge. One of these fields is bioinformatics [22], which aim is to understand biological data by extracting lengthy gene expressions that can be represented as colossal patterns [40].

Given a set of items $I = \{i_1, i_2, \ldots, i_n\}$ in a dataset, a colossal pattern P is formally defined as a subset of I, i.e. $\{P = \{i_j, \ldots, i_k\} \subseteq I, 1 \leq j, k \leq n\}$, which length $|P|$ is too high, and any of its sub-patterns has a similar frequency—there is no high drop in frequency when some items are added. For instance, an item i_1 that is present in a dataset with a frequency of 200, i.e. $f(i_1) = 200$, might experiment a drop in the frequency when it is extended with item i_2 and its frequency is $f(\{i_1, i_2\}) = 195$. If a new pattern i_3 is added, then the frequency drops to 193. The process continues till a large itemset with similar frequency is obtained, e.g. $P = \{i_1, i_2, \ldots i_{100}\}$ having a frequency of $f(P) = 186$. If we add a new item i_n and it produces a significant drop in the frequency, e.g. $P = \{i_1, i_2, \ldots i_{100}, i_n\}$ having $f(P) = 170$, then it cannot be added and the pattern $P = \{i_1, i_2, \ldots i_{100}\}$ is defined as colossal.

The definition of colossal patterns also introduces the term *core pattern*, which is really useful to fully understand how colossal patterns can be obtained. Given a pattern P defined as colossal, which comprises a set of sub-patterns sharing a similar frequency, we define a pattern P' as core if it is a sub-pattern of P, i.e. $P' \subset P$, and $f(P') \approx f(P)$. This similarity in frequency is defined by means of the core ratio τ, which determines that $f(P)/f(P') = \tau, 0 < \tau \leq 1$. A major feature of any colossal pattern is its robustness in the sense that it tends to have a high number of core patterns.

It should be noted that the process of computing all its frequent sub-patterns is computationally hard. For example, having the aforementioned colossal pattern $P = \{i_1, i_2, \ldots i_{100}\}$, which comprises 100 different items, i.e. $|P| = 100$, the number of frequent sub-patterns that may need to be computed is $2^{100} - 2$. At this point, the use of closed and maximal frequent patterns may partially alleviate this computational problem. Nevertheless, a colossal pattern cannot be compared with closed/maximal frequent patterns. First, every colossal pattern discovered in a dataset can also be defined as a closed frequent pattern, but every closed frequent pattern cannot necessarily be a colossal pattern. Second, a colossal pattern is not necessarily a maximal frequent pattern, e.g. considering the aforementioned example where item i_{101} is added, the resulting pattern can be frequent and colossal, so the previous pattern $P = \{i_1, i_2, \ldots i_{100}\}$ cannot be a maximal frequent pattern since one of its supersets is frequent. Thus, the problem of mining colossal patterns is considered as a different and challenging problem addressed by many different authors [31, 40].

1.2.6 Sequential Patterns

In previous sections, we have considered the pattern mining task as the extraction of strongly related items that appear together in a dataset. In some situations, a pattern might be considered of high interest not only due to the strong relationship between the items that it comprises but also due to the sequence of items [21] followed to form the pattern [1]. For example, consider the sequence of events $S = \langle \{Orange\}, \{Bread, Butter\}, \{Milk, Banana\} \rangle$, which is illustrated in Fig. 1.5. It describes that customers typically buy oranges, then bread and butter (not matter the temporal order) and, then milk and bananas (not matter the temporal order). Thus, each sequence S is described as a set of events $S = \langle e_1 \rightarrow \ldots \rightarrow e_n \rangle$ and each even e_i is described as an itemset $\{i_i, \ldots, i_j\}$ defined in the set of items I of a dataset [23]. The different events appear in a temporal order, so considering the example illustrated in Fig. 1.5, it is obtained that the item $\{Orange\}$ was bought earlier than the set of items $\{Butter, Cake\}$.

Formally, let us consider a set of items $I = \{i_1, i_2, \ldots, i_n\}$. An event e_j is defined as a non-empty unordered collection of items, i.e. $\{e_j = \{i_k, \ldots, i_m\} \subseteq I, 1 \leq k, m \leq n\}$. Additionally, it is possible to define a sequence S as an ordered list of events, i.e. $S = \langle e_1, e_2, \ldots, e_l \rangle$. The length of the sequence is defined as the sum of the lengths (number of individual items) of the different events, and a sequence

Fig. 1.5 Sequence of events for a sample market basket customer

con k items is defined as a k-sequence, and k is defined as the number of items included in the sequence, i.e. $k = \sum_{j=1}^{l} |e_j|$. As a matter of example, let us consider the sequence $S = \langle \{Orange\}, \{Bread, Butter\}, \{Milk, Banana\} \rangle$ shown in Fig. 1.5, which can be described as a 5-sequence that comprises three different events.

Given a sequence S, another sequence S' is defined as a sub-sequence of S if and only if there exist an event in S' that is a subset of S, and the events should keep ordered. As a matter of example, let as consider the aforementioned sequence S, which was defined as $S = \langle \{Orange\}, \{Bread, Butter\}, \{Milk, Banana\} \rangle$, the sequence $S' = \langle \{Bread\}, \{Milk, Banana\} \rangle$ is a sub-sequence of S, i.e. $S' \subset S$, since $e_{S_1'} \subset e_{S_2} \wedge e_{S_2'} \subset e_{S_3}$. On the contrary, the sequence $S'' = \langle \{Orange\}, \{Bread, Banana\} \rangle$ is not a sub-sequence of S since $e_{S_2''} \not\subset e_{S_2} \wedge e_{S_2''} \not\subset e_{S_3}$, i.e. despite the fact that the first even of S' is a subset of S, the second event of S' is not a subset of any of the remain events of S.

1.2.7   Spatio-Temporal Patterns

In the previous section we described sequential patterns as a way of dealing with some kind of features that introduce the time stamp in some way, for instance, to describe sequential behaviour [21]. The use of the time stamp can also be dealt in conjunction with spacial features, and this synergy gives rise to the term spatio-temporal patterns [8]. The concept of spatio-temporal patterns is related to those patterns that represent spatial and temporal information.

Nowadays, there is a high interest in the development of positioning technologies, which has caused a huge quantity of data collected from different devices. In many occasions, these devices act as real-time monitoring systems that produce high quantities of raw data that is meaningless without an in depth analysis [1]. For example, the movement of a specific object can be described by means of a sequence of spatial locations measured at different and consecutive time stamps. Data gathered by capturing the movement of this object may be of interest to discover frequently repeated paths. This process of mining spatio-temporal patterns might help to analyse past movements of the object in order to understand future movements.

Given a set of n locations described by the coordinates (x_i, y_i), and a specific time stamp t_i, then a spatio-temporal sequence S is defined as a list of locations ordered by the time stamp, i.e. $S = \langle \{(x_1, y_1), t_1\}, \{(x_2, y_2), t_2\}, \ldots, \{(x_n, y_n), t_n\} \rangle$. This type of patterns is highly related to sequential patterns, where the set of items described in each event represents locations defined by the coordinates (x_i, y_i).

1.3 Pattern Space Pruning

The task of finding all patterns in a database is quite challenging since the search space exponentially increases with the number of single items occurring in the database. Given a dataset comprising n singletons, a maximum number of $2^n - 1$ different patterns can be found in that dataset. Additionally, this dataset might contain plenty of transactions and the process of calculating the frequency for each pattern might be considered as a tough problem [10]. Let us consider a dataset comprising $n = 50$ singletons, and $m = 100,000$ transactions, the process of calculating the frequency of each pattern requires $(2^n - 1) \times m \times n$ different comparisons, so for the aforementioned example, the process requires $(2^{50} - 1) \times 100,000 \times 50 = 5.62910^{21}$ comparisons.

The required computational cost is too high when the datasets used in the mining process become bigger and bigger, which has highlighted the importance of pruning the search space, given rise to the paradigm of constraint-based mining [1]. Users are allowed to express their focus in mining by means of a set of constraints that capture application semantics. It helps users' exploration and control, and the paradigm allows to confine the search space of patterns to those of interest to the users achieving superior performance.

Constraints are really useful in pattern mining since it allows to prune the search space, thus reducing time and resources requirements. The use of constraints in pattern mining can be considered as an optimization problem where the use of constraints can be applied to the set of all patterns to test the constraint satisfaction. Nevertheless, this is not an optimal way of using constraints, and many efficient approaches [17] analyse the properties of the constraints and compute the patterns based on these constraints [25].

The use of constraints in pattern mining provides the user with a way of focussing on the interesting knowledge. It should be noted that these constraints enables a lower number of patterns to be obtained, but the set of patterns discovered is much more interesting. In pattern mining, constraints cannot be categorized into a fixed group and many different categories of constraints can be deemed. For example, for experts into a specific domain, it can be of interest to consider some constraints from the application domain's point of view. Let us consider the analysis of the market basket considering the set of items I, where the manager is interested in extracting knowledge about a group G of products related to babies. In such a way, an item constraint is required, which is defined as a constraint that denotes the specific group of items $G \subset I$ that can appear or cannot appear in any pattern P. Thus, the mining process is carried out by focusing only on those patterns that include specific items.

An additional constraint based on the application domain is a constraint related to the length of the patterns [19]. In some domains, it is of interest the mining of short patterns to increase their comprehensibility without mining too small patterns. Thus, a constraint that guides the length of the patterns is quite interesting.

Alternatively, constraints can also be categorized by means of their mining properties [12]. Two of the most well-known constraints in this sense are anti-monotone and monotone. A constraint C is defined as anti-monotone if and only if whenever a pattern P violates C, so does any superset $P_{sup} \supseteq P$, i.e. $\{C(P) = false \Rightarrow \forall P_{sup} \supseteq P : C(P_{sup}) = false\}$. At the same time, whenever a pattern P does not violate C, so does not any subset $P_{sub} \subseteq P$, i.e. $\{C(P) = true \Rightarrow \forall P_{sub} \subseteq P : C(P_{sub}) = true\}$. An example of anti-monotone constraint is the frequency of occurrence of a pattern (see Fig. 1.6). Thus, for a frequent pattern P, considering a minimum frequency threshold $f(P) = 2$, any of its subsets $P_{sub} \subseteq P$ is also frequent; whereas for an infrequent pattern P none of its supersets $P_{sup} \supseteq P$ is frequent (see Fig. 1.6). Considering the frequent pattern $P = \{Banana, Butter, Milk\}$, any sub-pattern $P_{sub} \subseteq P$ is also frequent: $P_1 = \{Banana, Butter\}$, $P_2 = \{Banana, Milk\}$,

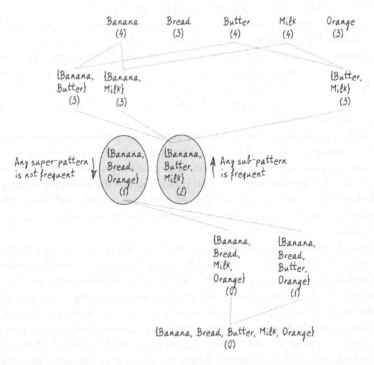

Fig. 1.6 Sub-lattice of the sample market basket that shows the anti-monotone property, i.e. any sub-pattern of a frequent pattern is also frequent, and any super-pattern of an infrequent pattern is never frequent

$P_3 = \{Butter, Milk\}$, $P_4 = \{Banana\}$, $P_5 = \{Butter\}$, and $P_6 = \{Milk\}$. Given a frequent pattern P having 3 singletons, i.e. $|P| = 3$, it contains $2^3 - 2 = 6$ sub-patterns that are also frequent. Considering now the infrequent pattern $P = \{Banana, Bread, Orange\}$, any super-pattern $P_{sup} \supseteq P$ is also frequent (see Fig. 1.6): $P_1 = \{Banana, Bread, Milk, Orange\}$, $P_2 = \{Banana, Bread, Butter, Orange\}$, and $P_3 = \{Banana, Bread, Butter, Milk, Orange\}$. Finally, a constraint C is defined as monotone if and only if whenever a pattern P satisfies C, so does any superset $P_{sup} \supseteq P$.

1.4 Traditional Approaches for Pattern Mining

The task of mining frequent patterns is one of the most well-known and intensively researched problems in data mining [12]. Numerous algorithms have been designed and developed to solve problems related to computational and memory requirements. In this regard, different frameworks have been described to solve the pattern mining problem, but most of them are based on a support framework [30], in which patterns with a frequency above a predefined threshold are extracted.

The first algorithm for mining frequent patterns is known as Apriori [5]. This algorithm is based on a level-wise paradigm in which all the frequent patterns of length $k + 1$ are generated by using all the frequent patterns of length k. Thus, the main characteristic of Apriori is that every subset of a frequent pattern is also frequent, so it follows the anti-monotone property, i.e. if a pattern $P = \{i_1, i_2\}$ is frequent, both i_1 and i_2 should be also frequent. In order to obtain new patterns of length $k + 1$, the algorithm uses joins of frequent patterns of length k as detailed in Fig. 1.7. A join is defined as pairs of frequent k-patterns that have at least $k - 1$ items in common. It should be noted that the same candidate can be produced by joining multiple frequent patterns.

Require: I, T, thr {items, transactions and minimum frequency threshold}
Ensure: L
1: $L_1 = \{\forall i | i \in I \wedge f(i) \geq thr\}$
2: $C = \emptyset$
3: **for** $(k = 1; L_k \neq \emptyset; k++)$ **do**
4:      $C =$ candidates patterns generated from L_k
5:      **for all** $t \in T$ **do**
6:          increment the frequency of all candidates in C contained in t
7:          $L_{k+1} =$ candidates in C that satisfies minimum frequency threshold trh,
           i.e. $L_{k+1} = \{\forall c \in C | f(c) \geq thr\}$
8:      **end for**
9: **end for**
10: **return** $L = \cup_k L_k$

Fig. 1.7 Pseudo-code of the Apriori algorithm

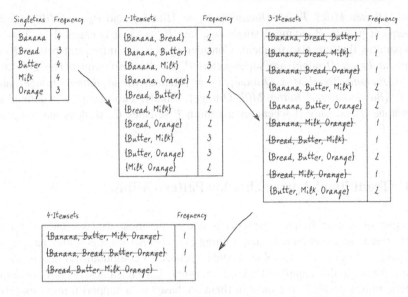

Fig. 1.8 Sample run of the Apriori algorithm illustrating the generation of the k-frequent pattern

One of the major features of Apriori is the use of a pruning strategy based on the frequency of occurrence of each specific itemset. This pruning strategy enables the exponential growth of candidate patterns to be controlled. For a better understanding, Fig. 1.8 illustrates a sample running of Apriori for the set of transactions shown in Table 1.1. Similarly to the previous examples, we assume that the minimum frequency threshold is $f_{min} = 2$, i.e. a pattern P is frequent if $f(P) \geq 2$. As previously described, a candidate $(k + 1)$-pattern is generated by joining two frequent patterns of size k which have at least $k-1$ items in common. For example, the frequent patterns $P_1 = \{Banana, Bread\}$ and $P_2 = \{Banana, Butter\}$ are joined to form a pattern of size three, i.e. $\{P' = \{Banana, Bread, Butter\} : P' \supset P_1 \wedge P' \supset P_2\}$. In the generation of the set of 3-itemsets, some of these 3-itemsets are discarded since they do not satisfy the frequency constraint. For example, the pattern $P' = \{Banana, Bread, Butter\}$ does not satisfy the minimum frequency threshold since $f(P') = 1 < f_{min} = 2$. Finally, it should be noted that the number of candidates k-itemsets is equal to $\binom{j}{k}$, j defined as the number of different singletons. Thus, the number of candidate patterns in C_3 is equal to $\binom{5}{3} = 10$, whereas L_3 is formed by the frequent patterns so $|C_k| \geq |L_k|$. Thus, considering a brute force approach, the set of 4-itemsets comprises $\binom{5}{4} = 5$ different patterns, whereas this number is lower if we consider the Apriori pruning, which obtains three itemsets of length four. It should be noted that the set of candidate itemsets C_{k+1} is obtained from the set of frequent patterns L_k obtained in the previous iteration.

An alternative procedure for the generation of candidate sets is to obtain $(k + 1)$-patterns by joining k-patterns with frequent singletons. In this procedure, it is not required to join two frequent patterns to form a new pattern, e.g.

$P_1 = \{Banana, Bread\} \cup P_2 = \{Banana, Butter\} \rightarrow P' = \{Banana, Bread, Butter\}$.
Instead, $P_1 = \{Banana, Bread\}$ is extended with the frequent singleton *Butter*,
giving rise to the candidate pattern $P' = \{Banana, Bread, Butter\}$. Nevertheless, this
procedure is still computationally complex when the number of transactions is too
high, since the frequency counting requires to analyse the complete dataset for each
different candidate set. To alleviate this issue, it is possible to carry out a transaction
reduction. A transaction that does not satisfy any frequent pattern should be marked
as useless in subsequent scans, so this transaction can be removed to reduce the
complexity. As a matter of example, let us consider the dataset shown in Table 1.1,
and the sample run of the Apriori algorithm illustrated in Fig. 1.8. In the generation
of the set of 3-itemsets, some of these itemsets are discarded since they do not satisfy
the frequency constraint, e.g. $P = \{Banana, Bread, Milk\}$. This infrequent itemset
can be removed from the dataset since it forms a whole transaction, so the number of
transactions can be reduced in order to reduce the complexity of the mining process
(see Fig. 1.9). Thus, in the process of computing the frequencies for the candidate
set C_4, the number of transactions is reduced, and hence, the required computational
complexity.

The reduction of the number of transactions in the dataset does not guarantee
that the computational complexity is highly reduced. In fact, in many occasions
the number of transactions cannot be reduced since no transaction represents an
infrequent patterns. In this regard, some authors [29] have considered the reduction
of the complexity of computing the frequencies of the items by splitting the original
database into m different sub-databases. The Apriori algorithm is run on each
sub-database separately, and the mined itemsets are combined for the different subdatasets. This way of mining patterns enables the number of transactions to be
reduced, so the process for computing the frequencies is computationally much less

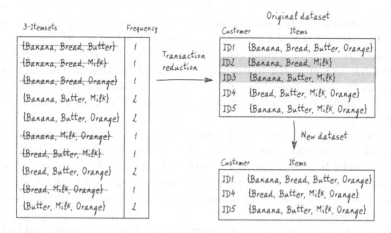

Fig. 1.9 Sample transaction reduction by removing transactions according to the set of infrequent
itemsets discovered in C_3

Table 1.6 Vertical data representation obtained from the market basket dataset shown in Table 1.1

Items	List
Banana	$\{ID_1, ID_2, ID_3, ID_5\}$
Bread	$\{ID_1, ID_2, ID_4\}$
Butter	$\{ID_1, ID_3, ID_4, ID_5\}$
Milk	$\{ID_2, ID_3, ID_4, ID_5\}$
Orange	$\{ID_1, ID_4, ID_5\}$

expensive. It should be noted that any itemset that is potentially frequent (according to a relative frequency) in the original database must be frequent in at least one of the partitions of this database.

As previously described, Apriori-based algorithms discover candidate itemsets in a breath-first search manner, k-itemsets are processed before $(k + 1)$-itemsets. Zaki [37] also proposed a breath-first algorithm, Equivalence CLAss Transformation (ECLAT), which transforms the original dataset into a vertical database format. Each single item is stored in the dataset together with a list of transaction-ids where the item can be found. For example, considering the sample dataset using along this chapter (see Table 1.1), a new vertical dataset is obtained (see Table 1.6). Here, each transaction t that describes a sample item, e.g. *Banana*, is denoted as $t(Banana) = \{ID_1, ID_2, ID_3, ID_5\}$. Two interesting properties should be highlighted for the set of transaction-ids. First, $t(X) = t(Y)$ if and only if X and Y happen together. For example, $t(Bread, Butter) = t(Bread, Orange)$, since $t(Bread, Butter) = \{ID_1, ID_4\}$ and $t(Bread, Orange) = \{ID_1, ID_4\}$. Second, $t(X) \subset t(Y)$ if and only if X is a subset of Y, e.g. $t(Bread, Butter) = t(Orange)$, since $t(Bread, Butter) = \{ID_1, ID_4\}$ and $t(Orange) = \{ID_1, ID_4, ID_5\}$.

ECLAT considers the frequency of a pattern P as the length of the transaction-ids list, so this algorithm determines that any itemset is frequent if it lists at least f_{min} transaction-ids, i.e. $|t(P)| \geq f_{min}$. The algorithm obtains new itemsets by using recursive intersections of candidate itemsets. For example, the item *Banana* comprises the transaction-id list $t(Banana) = \{ID_1, ID_2, ID_3, ID_5\}$, whereas the item *Bread* comprises the transaction-id list $t(Bread) = \{ID_1, ID_2, ID_4\}$. The intersection of these two lists gives rise to the frequent pattern $P = \{Banana, Bread\}$, which comprises the transaction-id list $t(P) = \{ID_1, ID_2\}$. Because every singleton is frequent in Table 1.6, there are 10 different intersections performed in total as shown in Table 1.7.

An important feature of ECLAT is the use of the anti-monotone property. Let us consider the infrequent pattern $P = \{Banana, Bread, Milk\}$, none of its super-itemsets can be frequent since $t(P) = \{ID_2\}$, i.e. $|t(P)| = 1 < f_{min} = 2$, and the intersection of this list with another cannot produce a frequent list. Finally, it should be noted that all this transaction-id lists can be keep en memory and they can be processed independently.

Another well-known method for mining frequent itemsets was proposed by Han et al. [13], which is called FP-Growth. A major feature of this algorithm is its ability to reduce the database scans by considering a compressed representation of the

Table 1.7 Vertical data representation for the set of 2-itemsets

Itemsets	List
{Banana, Bread}	{ID_1, ID_2}
{Banana, Butter}	{ID_1, ID_3, ID_5}
{Banana, Milk}	{ID_2, ID_3, ID_5}
{Banana, Orange}	{ID_1, ID_5}
{Bread, Butter}	{ID_1, ID_4}
{Bread, Milk}	{ID_2, ID_4}
{Bread, Orange}	{ID_1, ID_4}
{Butter, Milk}	{ID_3, ID_4, ID_5}
{Butter, Orange}	{ID_1, ID_4, ID_5}
{Milk, Orange}	{ID_4, ID_5}

database thanks to a data structure known as FP-Tree [36]. This structure is based on a traditional prefix-tree, which represents each node by using an item and the frequency of the itemset denoted by the path from the root to that node.

In order to construct the FP-Tree, we start with the empty node *null* and the database is scanned for each transaction. In the next step, the first transaction is analysed and every item is added to the root *null* so a branch of the tree is constructed. The process will continue with every transaction and the relevant nodes that are affected by the new insertion are incremented by one. If a new transaction shares the items of any branch of the tree, then the inserted transaction will be in the same path from the root to the common prefix. Otherwise, new nodes are inserted in the tree by creating new branches, with support count initialized to one. This process ends when all transactions have been inserted.

Considering the sample market basket dataset (see Table 1.1), the FP-Growth algorithm analyses each transaction in order to construct the FP-Tree structure (see Fig. 1.10). Thus, each transaction within the database is mapped onto a path in the tree. The first transaction includes the items {*Banana, Bread, Butter, Orange*}, so this string of items creates a new branch with all its nodes initialized to the value 1. In a second iteration, the second transaction shares the items {*Banana, Bread*} of the existing branch, so shares nodes will be incremented by one, and different items will form a new sub-branch. This process continues till all the transactions are analysed. It should be noted that one of the most important things to be deemed in this algorithm is the use of a fixed order among the items. Otherwise, the FP-Tree structure obtained by this algorithm is meaningless. Once the FP-Tree has been constructed, no further passes over the dataset are necessary. Any frequent itemset can be obtained directly from the FP-Tree structure by exploring the tree from the bottom-up. For example, the frequency of the pattern $P = \{Banana, Butter, Milk\}$ is $f(P) = 2$ according to the FP-Tree structure (see Fig. 1.11), and no analysis on the dataset is required.

A study on the performance of the FP-Growth algorithm has determined that it is efficient and scalable, and it is about an order of magnitude faster than Apriori. However, FP-Growth is not without problems, and the construction and traversing

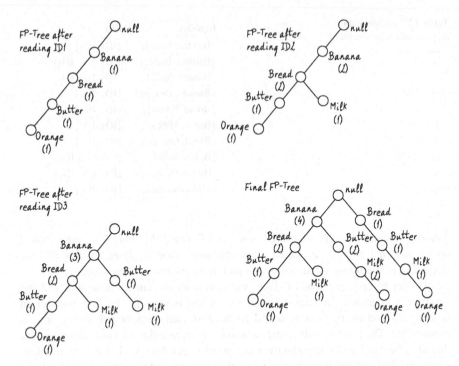

Fig. 1.10 Construction of the FP-Tree structure by using the sample market basket dataset shown in Table 1.1

Fig. 1.11 Frequency of occurrence of the pattern $P = \{Banana, Butter, Milk\}$ is 2. *Bold nodes* represents the path to obtain the frequency of this pattern

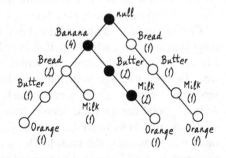

of the FP-Tree is not trivial. Besides, this algorithm requires both top-down and bottom-up traversals of the FP-Tree. To construct the tree, this algorithm requires to traverse the tree from top to down, but the process of mining frequent patterns is in the opposite way. Zhenguo et al. [39] have presented an approach that improves FP-Growth both in runtime, and also in the memory consumption. This algorithm, known as IFP-Growth, is based on a compound single linked list, which is used to improve the structure of the FP-Tree.

Back to the sample market basket dataset (see Table 1.1), IFP-Growth makes a first analysis on the database to obtain all the frequencies of the singletons, which will denote the first element of the linked lists. After this first process, the algorithm

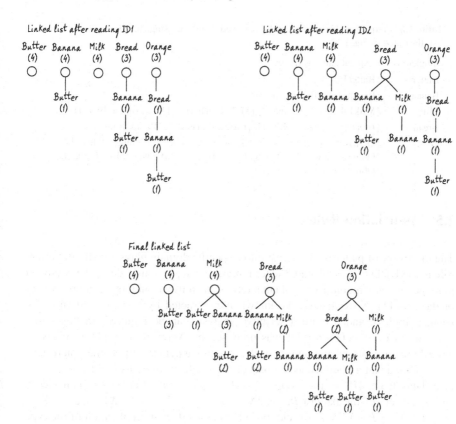

Fig. 1.12 Construction of the linked list by using the sample market basket dataset illustrated in Table 1.1

analyse each feasible itemset for each transaction and they will be linked to the corresponding singletons. For example, the first transaction will produce the linked list shown in Fig. 1.12. When the second transaction is analysed, new links are added to existing singletons, since {*Milk, Bread*} appears as an itemset for this transactions. Existing links will increase its frequency in one unity, and this process will continue till al the transactions are analysed, obtaining a final linked list (see Fig. 1.12).

Once the linked list has been created, the algorithm look for frequent patterns within the list. In this regard, IFP-Growth analyses all the lists by considering a minimum frequency threshold, e.g. $f_{min} = 2$. The set of all the frequent patterns that can be mined for the linked list is shown in Table 1.8. A major feature of this algorithm is that it does not require both top-down and bottom-up traversals of the lists, instead, just a top-down traversal is enough to obtain different itemsets. Finally, it should be noted that, similarly to FP-Growth, items should be placed in lexicographical order so the linked list used will depend on the first item to be analysed.

Table 1.8 Frequent patterns that can be obtained from each singleton by using the linked list illustrated in Fig. 1.12

Singletons	Frequent patterns:frequency
Butter	{Butter}:4
Banana	{Banana}:4, {Banana, Butter}:3
Milk	{Milk}:4, {Milk, Butter}:3, {Milk, Banana}:3, {Milk, Butter, Banana}:2
Bread	{Bread}:3, {Bread, Milk}:2, {Bread, Banana}:2, {Bread, Butter}:2
Orange	{Orange}:3, {Orange, Bread}:2, {Orange, Milk}:2, {Orange, Banana}:2, {Orange, Butter}:3, {Orange, Bread, Butter}:2, {Orange, Milk, Butter}:2, {Orange, Banana, Butter}:2,

1.5 Association Rules

The discovery of patterns of interest into a specific domain plays a really important role in the KDD process. Under the term *pattern*, we previously defined any type of homogeneity and regularity in data that represents intrinsic and important properties of datasets [1]. Nevertheless, the knowledge extracted by a single pattern might be meaningless, and a more descriptive analysis can be required. In this sense, the concept of association rules was proposed by Agrawal et al. [4] as a way of describing correlations among sets of items within a pattern of potential interest.

Let P be a pattern defined as a subset of the whole set of items $I = \{i_1, i_2, \ldots, i_n\}$ in a dataset, i.e. $\{P = \{i_j, \ldots, i_k\} \subseteq I, 1 \leq j, k \leq n\}$. Let us also consider X and Y subsets of the pattern P, i.e. $\{X \subset P \subseteq I \wedge Y = P \setminus X\}$, or also $\{Y \subset P \subseteq I \wedge X = P \setminus Y\}$. An association rule is defined as an implication of the type $X \rightarrow Y$ where X and Y do not have any common item, i.e. $X \cap Y = \emptyset$ since $\{X = P \setminus Y \wedge Y = P \setminus X\}$. The left-hand side of any association rule is denoted as antecedent of the rule, whereas the right-hand side is defined as consequent [5].

The meaning of an association rule $X \rightarrow Y$ is that if the antecedent X is satisfied, then it is highly probable that the consequent Y is also satisfied, i.e. one of the sets leads to the presence of the other set [38]. In pattern mining, the description of this type of relationships between items is quite interesting, since it describes the probability of satisfying the consequent once it is known that the antecedent was already satisfied. Similarly to the pattern mining task, the discovery of association rules was originally designed for the market basket analysis to obtain relations between products like *butter* $\rightarrow$ *bread* that describes the high probability of someone buying *bread* once he has bought *butter* previously.

Since its introduction in early 1990s, more and more domains [17, 19, 24] have been described and analysed by using association rules. In this sense, the description of unknown relations in data has given rise to an interesting data mining technique, known as association rule mining. Let us consider be a dataset comprising a set of transactions T, and each single transaction $t_i \subset T$ having a set of items within I, i.e. $t_i \subseteq I$. Association rule mining might be defined as the task of finding all association rules (AR) that satisfy specific user thresholds [7], i.e., $\{\forall \ AR \ | \ quality(AR) \geq q_{threshold}\}$.

The process of mining association rules is a really challenging task that suffers from different main problems that have been studied by many researchers [5, 38]. First proposals for mining association rules discover association rules by following two different steps. They first discover all existing frequent patterns, and then, this set of frequent patterns is used to discover any existing association rule that satisfies some specific user thresholds or quality values. Given a frequent pattern P that comprises a set of items, i.e. $\{P \subset I : P = \{i_j, \ldots, i_k\}, i \geq 1 \wedge k \leq |I|\}$, it can produce up to $2^n - 2$ different association rules, n defined as the length $|P|$ of the pattern. Noted that those rules obtained from P should be ignored if they have either an empty antecedent or consequent should be ignored, i.e. $\{X \rightarrow Y : X = \emptyset \wedge Y = P\}$ or $\{X \rightarrow Y : X = P \wedge Y = \emptyset\}$. As a matter of example, let $P = \{Banana, Butter, Milk\}$ be a frequent pattern. There are $2^{|P|} - 2 = 2^3 - 2 = 6$ different association rules generated from P:

- Rule 1: $\{Banana, Butter\} \rightarrow \{Milk\}$
- Rule 2: $\{Banana, Milk\} \rightarrow \{Butter\}$
- Rule 3: $\{Butter, Milk\} \rightarrow \{Banana\}$
- Rule 4: $\{Banana\} \rightarrow \{Butter, Milk\}$
- Rule 5: $\{Butter\} \rightarrow \{Banana, Milk\}$
- Rule 6: $\{Milk\} \rightarrow \{Banana, Butter\}$

Noted that rules comprising empty sets, i.e. $\{Banana, Butter, Milk\} \rightarrow \emptyset$ and $\emptyset \rightarrow \{Banana, Butter, Milk\}$, are meaningless since they do not provide any useful correlation between items.

One of the main problems of mining association rules is that its process is computationally expensive and requires a large amount of memory [20]. For a better understanding, let us consider a dataset containing k items, i.e. $|I| = k$. In this dataset, there are $\binom{k}{2}$ itemsets of length 2, and each of these itemsets can produce $2^2 - 2 = 2$ association rules. Considering itemsets of length 3, there are $\binom{k}{3}$ itemsets and each of them can produce $2^3 - 2 = 6$ association rules. This process is repeated till $\binom{k}{k}$, which produces $2^k - 2$ different association rules. Thus, in general, we can determines that for a dataset comprising k singletons, a total of $3^k - 2^{k+1} + 1$ different association rules might be generated.

Let us consider again the number of grains of sand on all the world's beaches, which was estimated to the value 10×10^{36}. Analysing the number of different rules that can be produced in a dataset comprising 150 singletons, we obtain a value of $3^{150} - 2^{151} + 1 \approx 3.70 \times 10^{71}$, so the number of rules that can be obtained from a dataset comprising 150 singletons is almost 35 order of magnitude higher than the number of grains of sand on all the world's beaches.

Traditional algorithms [5] reduce the search space by using the anti-monotone property and they look for rules that satisfy some minimum quality thresholds. Many of these algorithms firstly mine frequent patterns and they obtain rules from the set of patterns previously mined [13]. In such situations where the minimum frequency value used to determine whether a pattern is frequent or not is to low, the number of both candidate sets and rules increases considerably so the computational problem is a challenge. In addition, many patterns having a weak correlation might be produced so the number of useless rules can be high. On the contrary, in situations where the minimum frequency value is too high, some of the discovered patterns may discard items with low supports, which might be useful for the discovery of associations among rare or abnormal items [18]. For instance, in some medical domains, abnormal or unusual disease cannot be discarded just because they do not appear frequently. In fact, this type of items can be considered as much more important than frequent items, so the application domain plays an important role in this regard [27].

References

1. C. C. Aggarwal and J. Han. *Frequent Pattern Mining*. Springer International Publishing, 2014.
2. C. C. Aggarwal and P. S. Yu. A New Framework For Itemset Generation. In *In Proceedings of the 1998 Symposium on Principles of Database Systems*, pages 18–24, 1998.
3. C. C. Aggarwal, Y. Li, J. Wang, and J. Wang. Frequent Pattern Mining with Uncertain Data. In *Proceedings of the 15th ACM SIGKDD International Conference on Knowledge Discovery and Data Mining*, KDD '09, pages 29–38, Paris, France, 2009. ACM.
4. R. Agrawal and R. Srikant. Fast Algorithms for Mining Association Rules in Large Databases. In *Proceedings of the 20th International Conference on Very Large Data Bases*, VLDB '94, pages 487–499, San Francisco, CA, USA, 1994. Morgan Kaufmann Publishers Inc.
5. R. Agrawal, T. Imielinski, and A. N. Swami. Mining association rules between sets of items in large databases. In *Proceedings of the 1993 ACM SIGMOD International Conference on Management of Data*, SIGMOD Conference '93, pages 207–216, Washington, DC, USA, 1993.
6. Michael J. Berry and Gordon Linoff. *Data Mining Techniques: For Marketing, Sales, and Customer Support*. John Wiley & Sons, Inc., New York, NY, USA, 2011.
7. F. Berzal, I. Blanco, D. Sánchez, and M. A. Vila. Measuring the Accuracy and Interest of Association Rules: A new Framework. *Intelligent Data Analysis*, 6(3):221–235, 2002.
8. H. Cao, N. Mamoulis, and D. W. Cheung. Mining frequent spatio-temporal sequential patterns. In *Proceedings of the 5th IEEE International Conference on Data Mining*, ICDM '05, Houston, Texas, USA, 2005.
9. L. Geng and H. J. Hamilton. Interestingness Measures for Data Mining: A Survey. *ACM Computing Surveys*, 38, 2006.
10. B. Goethals. Survey on Frequent Pattern Mining. Technical report, Technical report, HIIT Basic Research Unit, Department of Computer Science, University of Helsinki, Finland, 2003.
11. M. Gorawski and P. Jureczek. Extensions for Continuous Pattern Mining. In *Proceedings of the 2011 International Conference on Intelligent Data Engineering and Automated Learning*, IDEAL 2011, pages 194–203, Norwich, UK, 2011.
12. J. Han and M. Kamber. *Data Mining: Concepts and Techniques*. Morgan Kaufmann, 2000.
13. J. Han, J. Pei, Y. Yin, and R. Mao. Mining Frequent Patterns without Candidate Generation: A Frequent-Pattern Tree Approach. *Data Mining and Knowledge Discovery*, 8:53–87, 2004.
14. J. Han, H. Cheng, D. Xin, and X. Yan. Frequent Pattern Mining: Current Status and Future Directions. *Data Mining Knowledge Discovery*, 15(1):55–86, 2007.
15. Y. S. Koh and N. Rountree. *Rare Association Rule Mining and Knowledge Discovery: Technologies for Infrequent and Critical Event Detection*. Information Science Reference, Hershey, New York, 2010.
16. Y. Li, A. Algarni, and N. Zhong. Mining Positive and Negative Patterns for Relevance Feature Discovery. In *Proceedings of the 16th ACM SIGKDD International Conference on Knowledge*

Discovery and Data Mining, KDD '10, pages 753–762, Washington, DC, USA, 2010. ACM.

17. J. M. Luna, J. R. Romero, and S. Ventura. Design and behavior study of a grammar-guided genetic programming algorithm for mining association rules. *Knowledge and Information Systems*, 32(1):53–76, 2012.

18. J. M. Luna, J. R. Romero, and S. Ventura. On the adaptability of G3PARM to the extraction of rare association rules. *Knowledge and Information Systems*, 38(2):391–418, 2014.

19. J. M. Luna, C. Romero, J. R. Romero, and S. Ventura. An Evolutionary Algorithm for the Discovery of Rare Class Association Rules in Learning Management Systems. *Applied Intelligence*, 42(3):501–513, 2015.

20. J. M. Luna, A. Cano, M. Pechenizkiy, and S. Ventura. Speeding-Up Association Rule Mining With Inverted Index Compression. *IEEE* Transactions on Cybernetics, pp(99):1–14, 2016.

21. N. R. Mabroukeh and C. I. Ezeife. A taxonomy of sequential pattern mining algorithms. *ACM Computing Surveys*, 43(1):1–41, 2010.

22. M. Martinez-Ballesteros, I. A. Nepomuceno-Chamorro, and J. C. Riquelme. Inferring gene-gene associations from quantitative association rules. In *Proceedings of the 11th International Conference on Intelligent Systems Designe and Applications*, ISDA 2011, pages 1241–1246, Cordoba, Spain, 2011.

23. C. H. Mooney and J. F. Roddick. Sequential pattern mining – approaches and algorithms. *ACM Computing Surveys*, 45(2):1–39, 2013.

24. N. Ordoñez, C. Ezquerra and C. Santana. Constraining and Summarizing Association Rules in Medical Data. *Knowledge and Information Systems*, 9, 2006.

25. J. Pei and J. Han. Constrained frequent pattern mining: A pattern-growth view. *ACM SIGKDD Explorations Newsletter*, 4(1):31–39, 2002.

26. C. Romero and S. Ventura. Educational data mining: a review of the state of the art. *IEEE Transactions on Systems, Man, and Cybernetics, Part C*, 40(6):601–618, 2010.

27. C. Romero, J. M. Luna, J. R. Romero, and S. Ventura. Mining Rare Association Rules from e-Learning Data. In *Proceedings of the 3rd International Conference on Educational Data Mining*, EDM 2010, pages 171–180, Pittsburgh, PA, USA, 2010.

28. D. Sánchez, J. M. Serrano, L. Cerda, and M. A. Vila. Association Rules Applied to Credit Card Fraud Detection. *Expert systems with applications*, (36):3630–3640, 2008.

29. A. Savasere, E. Omiecinski, and S. B. Navathe. An efficient algorithm for mining association rules in large databases. In *Proceedings of the 21th International Conference on Very Large Data Bases*, VLDB '95, pages 432–444, San Francisco, CA, USA, 1995.

30. T. Scheffer. Finding association rules that trade support optimally against confidence. In *Proceedings of the 5th European Conference of Principles and Practice of Knowledge Discovery in Databases*, PKDD 2001, pages 424–435, Freiburg, Germany, 2001.

31. M. K. Sohrabi and A. A. Barforoush. Efficient colossal pattern mining in high dimensional datasets. *Knowledge-Based Systems*, 33:41–52, 2012.

32. R. Srikant and R. Agrawal. Mining Quantitative Association Rules in Large Relational Tables. In *Proceedings of the 1996 ACM SIGMOD International Conference on Management of Data*, SIGMOD'96, Montreal, Quebec, Canada, 1996.

33. L. Szathmary, A. Napoli, and P. Valtchev. Towards rare itemset mining. In *Proceedings of the 19th IEEE International Conference on Tools with Artificial Intelligence*, ICTAI '07, pages 305–312, Patras, Greece, 2007.

34. L. Szathmary, P. Valtchev, and A. Napoli. Generating Rare Association Rules Using the Minimal Rare Itemsets Family. *International Journal of Software and Informatics*, 4(3): 219–238, 2010.

35. P. Tan and V. Kumar. Interestingness Measures for Association Patterns: A Perspective. In *Proceedings of the Workshop on Postprocessing in Machine Learning and Data Mining*, KDD '00, New York, USA, 2000.

36. P. N. Tan, M. Steinbach, and V. Kumar. *Introduction to Data Mining*. Addison Wesley, 2005.

37. M. J. Zaki. Scalable algorithms for association mining. *IEEE Transactions on Knowledge and Data Engineering*, 12(3):372–390, 2000.

38. C. Zhang and S. Zhang. *Association rule mining: models and algorithms*. Springer Berlin /
    Heidelberg, 2002.
39. D. Zhenguo, W. Qinqin, and D. Xianhua. An improved fp-growth algorithm based on
    compound single linked list. In *Proceedings of the 2009 Second International Conference
    on Information and Computing Science*, ICIC '09, pages 351–353, Washington, DC, USA,
    2009. IEEE Computer Society.
40. F. Zhu, X. Yan, J. Han, P. S. Yu, and H. Cheng. Mining colossal frequent patterns by
    core pattern fusion. In *Proceedings of the IEEE 23rd International Conference on Data
    Engineering*, ICDE 2007, pages 706–71, Istanbul, Turkey, 2007. IEEE.

Chapter 2
Quality Measures in Pattern Mining

Abstract In this chapter different quality measures to evaluate the interest of the patterns discovered in the mining process are described. Patterns represent major features of data so their interestingness should be accordingly quantified by considering metrics that determine how representative a specific pattern is for the dataset. Nevertheless, a pattern can also be of interest for a user despite the fact that this pattern does not describe useful and intrinsic properties of data. Thus, any quality measure can be divided into two main groups: objective and subjective quality measures. Whereas objective measures describe statistical properties of data, subjective quality measures take into account both the data properties and external knowledge provided by the expert in the application domain.

2.1 Introduction

Pattern mining is defined as the process of extracting patterns of special interest from raw data [1]. Generally, the user does not have any information about data so any knowledge extracted from that data is completely new and the interest of the mined patterns is hardly quantifiable. Sometimes, though, the user previously knows what type of knowledge is useful to be obtained from data, and it makes possible to quantify the level of interest of the patterns [15] discovered by different pattern mining algorithms.

In general, patterns represent major features of data so their interest should be quantified by considering metrics that determine how representative a specific pattern is for the dataset [5, 11]. Good metrics should choose and rank patterns based on their potential interest to the user. Nevertheless, many application domains require specific knowledge to be discovered and any expert in the domain needs to quantify how promising a pattern is [26]. All of this give rise to the division of pattern mining metrics into two different groups: objective and subjective metrics.

Objective metrics are usually defined from a statistical point of view and they provide structural properties of data [27]. Some of these metrics have been widely used to evaluate the interest of association patterns [21], determining that the stronger is the dependence relationship, the more interesting is the pattern. Support, confidence, lift, and leverage are some of these objective metrics, which are defined in terms of the frequency of occurrence of the patterns.

© Springer International Publishing Switzerland 2016

S. Ventura, J.M. Luna, *Pattern Mining with Evolutionary Algorithms*,

DOI 10.1007/978-3-319-33858-3_2

As for the subjective metrics, they usually incorporate some users' knowledge in the application field. Blöttcher et al. [7] described that any association pattern is considered interesting if it is either actionable or unexpected. Actionability of a relationship means that the user might act upon it to his own advantage. Additionally, the unexpectedness refers to how far the discovered pattern contradicts the user's knowledge about the domain. Thus, most of the approaches for quantifying the subjective interest of a pattern require comparisons of the discovered knowledge with regard to the user's knowledge.

In general, most of the existing proposals for mining patterns and associations between patterns follow an optimization based on objective quality measures. This huge attraction for this type of metrics lies in the fact that pattern mining mainly aims at discovering hidden and previously unknown knowledge from datasets [14], so there is no possibility to compare the extracted knowledge with the subjective knowledge provided by the expert. Besides, the knowledge of two different users into a specific field can differ greatly, which causes inaccuracy in the metrics. Thus, the values obtained for different subjective metrics cannot be properly compared.

Finally, it should be noted that both objective and subjective measures can be used to select interesting rules. First, objective metrics serves as a kind of filter to select a subgroup of potentially interesting patterns and association between patterns. Second, subjective metrics might be used as a second filter to keep only those patterns that are truly interesting for both the user and the application field.

2.2 Objective Interestingness Measures

As described in the previous chapter, among the objectives of KDD (Knowledge Discovery in Databases) the production of patterns of interest plays one of the most important roles. In general terms, a pattern is an entity that should be valid, new and comprehensive [9]. Nevertheless, the mining of unknown patterns in a database might produce a large amount of different patterns, which causes a hardly post-process for the end user who needs to analyse and study each pattern individually. Besides, a large percentage of this set of patterns may be uninteresting and useless, so the end user has to face two different problems: the quantity and the quality of the rules [13]. To solve this issue, different quality measures based on the analysis of the statistical properties of data have been proposed by different authors [5, 26].

Let us consider an association rule $X \rightarrow Y$ defined from a pattern $P \subseteq I = \{i_1, i_2, \ldots, i_k\}$ obtained from a dataset, where X and Y are subset of P with no item in common, i.e. $\{X \subset P \land Y \subset P : X \cap Y = \emptyset, P \setminus X = Y, P \setminus Y = X\}$. The absolute frequencies for any association rule $X \rightarrow Y$ comprising an antecedent X and a consequent Y can be tabulated as shown in Table 2.1. We note n as the total number of transactions in the dataset. We also define n_x and n_y as the number of transactions that satisfies X and Y, respectively. n_{xy} is defined as the number of transactions that satisfies both X and Y, i.e. the number of transactions that satisfies the pattern P in which the association rule was defined. Additionally, $n_{x\bar{y}}$ is defined as the number

Table 2.1 Absolute frequencies for the antecedent X and consequent Y of any association rule of the form $X \to Y$

	Y	$\overline{Y}$	Σ
X	n_{xy}	$n_{x\overline{y}}$	n_x
$\overline{X}$	$n_{\overline{x}y}$	$n_{\overline{x}\overline{y}}$	$n_{\overline{x}}$
Σ	n_y	$n_{\overline{y}}$	n

Table 2.2 Relative frequencies for the antecedent X and consequent Y of any association rule of the form $X \to Y$

	Y	$\overline{Y}$	Σ
X	p_{xy}	$p_{x\overline{y}}$	p_x
$\overline{X}$	$p_{\overline{x}y}$	$p_{\overline{x}\overline{y}}$	$p_{\overline{x}}$
Σ	p_y	$p_{\overline{y}}$	1

of transactions that satisfies X but not Y, i.e. $n_{x\overline{y}} = n_x - n_{xy}$. Finally, $n_{\overline{x}y}$ is defined as the number of transactions that satisfies Y but not X, i.e. $n_{\overline{x}y} = n_y - n_{xy}$.

All these values can also be represented by considering the relative frequencies rather than the absolute ones. Thus, given the antecedent X and the number of transactions that it satisfies n_x, it is possible to calculate its relative frequency as $p_x = n_x/n$. The relative frequency of X describes, in per unit basis, the percentage of transactions satisfied by X. Table 2.2 illustrates the relative frequencies for a sample association rule $X \to Y$ comprising an antecedent X and a consequent Y. Analysing the relative frequencies, it should be noted that a rule is useless or misleading if $p_{xy} = 0$ since it does not represent any transaction. It could be caused by two different situations: (1) the antecedent X (or the consequent Y) does not satisfies any transaction within the dataset, so both X and Y does not have any transaction satisfied in common, i.e. $P_{xy} = 0$. (2) either the antecedent X and the consequent Y satisfy up to 50 % of the transactions within the dataset, i.e. $p_x \leq 0.5$ and $p_y \leq 0.5$, but they do not satisfy any transaction in common. In case the sum of the probabilities is greater than 1, then P_{xy} is always greater than 0, so it is impossible to have a probability of 0 if $p_x + p_y > 1$. In fact, it should be noted that the maximum value of P_{xy} is equal to the minimum value among P_x and P_y, i.e. $P_{xy} \leq Min\{P_x, P_y\}$. Figure 2.1 illustrates this behaviour, describing that in cases where $P_x + P_y \leq 1$ the value of P_{xy} is defined in the range $[0, Min\{P_x, P_y\}]$. On the contrary, in those situations where $P_x + P_y > 1$ the value of P_{xy} is defined in the range $[P_x + P_y - 1, Min\{P_x, P_y\}]$. All of this led us to the conclusion that when both P_x and P_y are maximum, i.e. $P_x = P_y = 1$, then P_{xy} is equal to 1.

The behaviour of both p_x and p_y with regard to p_{xy} is described by the equation $2 \times p_{xy} \leq p_x + p_y \leq 1 + p_{xy}$, which is graphically illustrated in Fig. 2.2. As shown, $P_{xy} > 0.5$ if and only if $P_x + Py > 1$. Considering the aforementioned equation, i.e. $P_{xy} \leq Min\{P_x, P_y\}$, P_{xy} will have a value greater or equal to 0.5 when $P_x + P_y \geq 1$.

As previously stated, P_{xy} is defined as the probability of occurrence of a pattern $P = \{X, Y\}$ that represent an association rule $X \to Y$ to be satisfied in a dataset. Here, P_x stands for the probability or relative frequency of the antecedent X of the rule, whereas P_y describes the relative frequency of the consequent Y of the rule. In pattern mining, this frequency of occurrence is widely known as the support [3] of

Fig. 2.1 P_{xy} may be 0 in
situations where $P_x + P_y \leq 1$.
On the contrary (example on
the right), $P_{xy} > 0$ in
situations where $P_x + P_y > 1$,
taking a minimum value of
$P_x + P_y - 1$

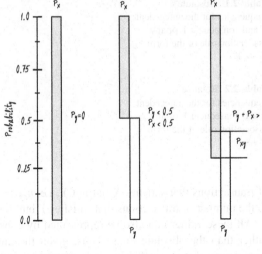

Fig. 2.2 Relationship
between the probability of a
rule P_{xy} and the sum of the
probabilities of both the
antecedent P_x and
consequent P_y

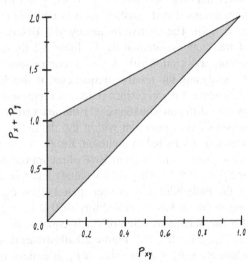

the pattern, being one of the major quality measures used in this field. Hence, given
an association rule of the form $X \rightarrow Y$, we define its frequency of occurrence as
$support(X \rightarrow Y) \equiv P_{xy}$.

2.2.1 Quality Properties of a Measure

In 1991, Piatetsky-Shapiro [20] suggested that any quality measure $\mathcal{M}$ defined to
quantify the interest of an association within a pattern should verify three specific
properties in order to separate strong and weak rules so high and low values can be
assigned, respectively. These properties can be described as follows:

Table 2.3 Properties satisfied by a set of different objective quality measures

Measure	Equation	Range	Property 1	Property 2	Property 3
Support	P_{xy}	[0, 1]	No	Yes	No
Coverage	P_x	[0, 1]	No	No	No
Confidence	P_{xy}/P_x	[0, 1]	No	Yes	No
Lift	$P_{xy}/(P_x \times P_y)$	[0, n]	Yes[a]	Yes	Yes
Leverage	$P_{xy} - (P_x \times P_y)$	[-0.25, 0.25]	Yes	Yes	Yes
Cosine	$P_{xy}/\sqrt{(P_x \times P_y)}$	[0, 1]	No	Yes	Yes
Conviction	$(P_x \times P_{\bar{y}})/P_{x\bar{y}}$	[0, ∞)	No	Yes	No
Gain	$(P_{xy}/P_x) - P_y$	[-1, 1)	Yes	Yes	Yes
Certainty factor	$((P_{xy}/P_x) - P_y)/(1 - P_y)$	[-1, 1]	Yes	Yes	Yes

[a]This property is satisfied just in case that the measure will be normalized

- Property 1: $\mathscr{M}(X \to Y) = 0$ when $P_{xy} = P_x \times P_y$. This property claims that any quality measure $\mathscr{M}$ should test whether X and Y are statistically independent.
- Property 2: $\mathscr{M}(X \to Y)$ monotonically increases with P_{xy} when P_x and P_y remain the same.
- Property 3: $\mathscr{M}(X \to Y)$ monotonically decreases with P_x or with P_y when other parameters remain the same, i.e. P_{xy} and P_x or P_y remain unchanged.

Following with the analysis of objective quality measures, all of them will be analysed by considering the properties defined by *Piatetsky-Shapiro*. Table 2.3 summarizes all the quality measures describes in this section, which will be described in depth. Beginning with the support quality measure, it should be noted that this measure does not satisfy first property since support$(X \to Y) \neq 0$ when $P_{xy} = P_x \times P_y$. For example, if $P_{xy} = P_x = P_y = 1$ then $P_{xy} = P_x \times P_y$, and support$(X \to Y) = 1$, so the first property defined by *Piatetsky-Shapiro* is not satisfied. Similarly, this quality measure does not satisfy the third property since support$(X \to Y) = P_{xy}$ so support measure cannot monotonically decrease when P_{xy} remains unchanged. Finally, it is interesting to note that a general belief in pattern mining is that the greater the support, the better the pattern discovered in the mining process. Nevertheless, this assertion should be taken with a grain of salt as it is only true to some extent. Indeed, patterns that appear in all the transactions are misleading since they do not provide any new knowledge about the data properties.

Similarly to support, there is a metric that calculates the generality [2] of a rule based on the percentage of transactions satisfied by the antecedent P_x of a rule, also known as body of a rule. This measure, known as coverage, determines the comprehensiveness of a rule. If a pattern characterizes more information in the dataset, it tends to be much more interesting. Nevertheless, this quality measure does not satisfy any of the properties described by *Piatetsky-Shapiro* as shown in Table 2.3. Besides, it is highly related to the support metric so most of the proposals have considered support instead of coverage.

As well as support, confidence is a quality measure that appears in most of the problems where the mining of association between patterns is a dare [29].

This quality measure determines the reliability or strength of implication of the rule, so the higher its value, the more accurate the rule is. In a formal way, the confidence measure is defined as the proportion of transactions that satisfy both the antecedent X and consequent Y among those transactions that contain only the antecedent X. This quality measure can be formally expressed as $confidence(X \rightarrow Y) = P_{xy}/P_x$, or as an estimate of the conditional probability $P(Y|X)$.

Support and confidence are broadly conceived as the finest quality measures in quantifying the quality of association rules and, consequently, a great variety of proposals make use of them [23]. These proposals attempt to discover rules whose support and confidence values are greater than certain thresholds. Nevertheless, many authors have considered that the mere fact of exceeding these quality thresholds does not guarantee that the rules are interesting at all [5]. For instance, the support-confidence framework does not provide a test for capturing the correlation of two patterns. Let us consider a real example obtained in [8], which described the rule IF *past active duty in military* THEN *no service in Vietnam*. This rule is calculated with a very high accuracy, which is quantified by the confidence value of 0.90. Thus, the rule suggests that knowing that a person served in military we should believe that he or she did not serve in Vietnam with a probability of 90 %. However, the item *no service in Vietnam* has a support of 0.95, so the probability that a person did not serve in Vietnam decreases (from 0.95 to 0.90) when we know he or she served in military. All of this led us to determine that the rule is misleading since the previously known information is more accurate than the probability obtained when more descriptive information is added.

Analysing whether the confidence quality measure satisfies or not the three properties provided by Piatetsky-Shapiro [20], we obtain that it only satisfies the second property (see Table 2.3). Let us consider the following probabilities: $P_{xy} = 1/3$, $P_x = 1/2$, and $P_y = 2/3$. The first property proposed by *Piatetsky-Shapiro* determines that if $P_{xy} = P_x \times P_y$ is satisfied by a measure $\mathcal{M}$, then this measure should satisfies that $\mathcal{M}(X \rightarrow Y) = 0$. Considering the aforementioned probabilities, $P_{xy} = P_x \times P_y = 1/3$ so the confidence will satisfy the first property if and only if $confidence(X \rightarrow Y) = 0$. Nevertheless, computing this value, we obtain that $confidence(X \rightarrow Y) = P_{xy}/P_x = 2/3 \neq 0$ so the confidence metric does not satisfy the property number one. Continuing with the second property, it is trivial to demonstrate that $confidence(X \rightarrow Y) = P_{xy}/P_x$ monotonically increases with P_{xy} when P_x remains the same, so this second property is satisfied by the confidence metric. Finally, the third property is partially satisfied by confidence measure since this measure does not include P_y, so we cannot state that third property is satisfied by confidence measure.

Support and confidence have been widely used in the mining of associations between patterns [29], and they are still considered by numerous authors that apply association to specific application fields [18, 22]. These two metrics are related as shown in Fig. 2.3, the shaded area illustrates the feasible area in which any association rule can obtain the values for the support and confidence measures. In order to understand the existing relation between these two quality measures,

Fig. 2.3 Relationship between the support and the confidence measures

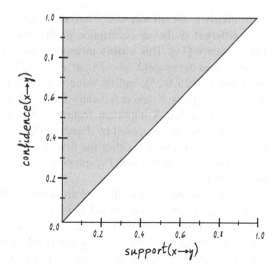

it should be noted that $P_{xy} \leq Min\{P_x, P_y\}$, and $P_{xy} \leq P_x$. Thus, given a value P_{xy}, then P_x will have a value in the range $P_x \in [P_{xy}, 1]$, so *confidence*$(X \rightarrow Y)$ is always greater or equal to *support*$(X \rightarrow Y)$.

Brin et al. [8] proposed a different metric to quantify the interest of the associations extracted in the mining process. This quality measure, which is known as lift, calculates the relationship between the confidence of the rule and the expected confidence of the consequent. Lift quality measure is described as *lift*$(X \rightarrow Y)= P_{xy}/(P_x \times P_Y) = confidence(X \rightarrow Y)/P_y$. This measure calculates the degree of dependence between the antecedent X and the consequent Y of an association rule, obtaining a value < 1 if they are negative dependent; a value > 1 if they are positive dependent; and 1 in case of independence. As shown in Table 2.3, if *lift*$(X \rightarrow Y) = 1$, then $P_{xy} = P_x \times P_y$ so X and Y are independent and this measure satisfies the first property of Piatetsky-Shapiro [20] just in case that the measure will be normalized; if *lift*$(X \rightarrow Y) > 1$, then this measure monotonically increases with P_{xy} when other parameters remain the same (Property 2 of *Piatetsky-Shapiro*); finally, if *lift*$(X \rightarrow Y) < 1$, then lift monotonically decreases with P_x (or P_y) when other parameters remain the same (Property 3 of *Piatetsky-Shapiro*).

Let us consider the same real example described for the confidence measure and proposed by Brin et al. [8]. This example determined that the rule IF *past active duty in military* THEN *no service in Vietnam* has a confidence value of 0.90, and the support of the pattern *no service in Vietnam* is 0.95. In this regard, considering the same rule, the lift measure obtains a value of *lift* $= 0.90/0.95 = 0.947$, describing a negative dependence between the antecedent and consequent. Thus, this quality measure describes much more information than the one obtained by the confidence metric. Finally, it should be noted that most authors look for a positive correlation among the antecedent X and the consequent Y so only values greater than 1 are desired, that is, the confidence of the rule is greater than the support of the consequent.

Similarly to the lift measure, *Piatetsky-Shapiro* proposed a metric that calculates how different is the co-occurrence of the antecedent and the consequent from independence [16]. This quality measure is known as novelty or leverage [4], and is defined as $leverage(X \rightarrow Y) = P_{xy} - (P_x \times P_y)$. Leverage takes values in the range $[-0.25, 0.25]$, and its value is zero in those cases where the antecedent X and consequent Y are statistically independent, so values close to zero imply uninteresting rules. A important feature of this quality measure is that it satisfies the three properties proposed by *Piatetsky-Shapiro*. First, $leverage(X \rightarrow Y) = 0$ if $P_{xy} = P_x \times P_y$, so it satisfies the first property. Additionally, $leverage(X \rightarrow Y)$ monotonically increases with P_{xy} (property 2), and monotonically decreases with P_x or with P_y (property 3).

Another quality metric derived from lift is the IS measure, also known as cosine, which is formally defined as $IS(X \rightarrow Y) = \sqrt{Lift \times P_{xy}}$. As described by the authors [26], the IS measure presents many desiderable properties despite it does not satisfies the first property described by *Piatetsky-Shapiro*. First, it takes into account both the interestingness and the significance of an association rule since it contains two important quality measures in this sense, i.e. support and lift. Second, IS is equivalent to the geometric mean of confidence, i.e. $IS = \sqrt{Lift \times P_{xy}} = \sqrt{P_{xy}^2/(P_x \times P_y)}$, which can also be described as $IS(X \rightarrow Y) = \sqrt{confidence(X \rightarrow Y) \times confidence(Y \rightarrow X)}$. Finally, this quality measure can be described as the cosine angle, i.e. $IS(X \rightarrow Y) = P_{xy}\sqrt{P_x \times P_y}$.

A major drawback of the lift quality measure is its bi-directional meaning, which determines that $lift(X \rightarrow Y) = lift(Y \rightarrow X)$ so it measures co-occurrence, not implication. It should be noted that the final aim of any association rule is the discovery and description of implications between the antecedent and consequent of the rule, so the direction of the implication is quite important and does not always reflect the same meaning and, therefore, cannot be measured with the same values. In this regard, the conviction quality measure (see Table 2.3) was proposed [8] as $conviction(X \rightarrow Y) = (P_x \times P_{\bar{y}})/P_{x\bar{y}}$. Conviction represents the degree of implication of a rule, and values far from the unity indicate interesting rules. According to the authors, this quality measure is useful for the following reasons:

- Conviction is related to both P_x and P_y, which is a great advantage with regard to confidence. As previously described, confidence only considers P_x so it may give rise to a confidence value that is smaller than the support of the consequent, i.e. a negative dependence between the antecedent and consequent.
- The value of this measure is always equal to 1 when the antecedent X and the consequent Y are completely unrelated. For example, a P_y value equal to P_{xy} produces a value of 1, which means independence between both X and Y.
- Unlike lift, rules which hold 100 % of the time have the highest possible conviction value. The confidence measure also has this property, providing a value of 1 (the maximum value for confidence) to these rules.
- Conviction is a measure of implication since it is directional, so it behaves differently to lift.

Considering the conviction quality measure, it should be noted that its main drawback lies in the fact that it is not a bounded measure, i.e. its range is $[0, \infty]$. This quality makes impossible to determine an optimum quality threshold, which is a really big handicap for the use of this metric.

A different quality measure that is based on both the confidence and the support of the consequent is defined as gain of a rule or relative accuracy [16]. This quality measure, which is formally defined as $gain(X \rightarrow Y) = confidence(X \rightarrow Y) - P_y$, satisfies the three properties described by *Piatetsky-Shapiro* as shown in Table 2.3. The gain measure can also be defined as $gain(X \rightarrow Y) = (P_{xy} - (P_x \times P_y))/P_x$ so when $P_{xy} = P_x \times P_y$, then $gain(X \rightarrow Y) = 0$. Thus, this metric satisfies the first property described provided by *Piatetsky-Shapiro*. Additionally, $gain(X \rightarrow Y)$ monotonically increases with P_{xy} (property 2), and monotonically decreases with P_x or P_y (property number 3).

Based on the same features provided by the gain measure, a different quality measure was defined in [24] as the gain normalized into the interval $[-1,1]$. This quality measure, known as certainty factor (CF), calculates the variation of the probability P_Y of the consequent of a rule when consider only those transactions satisfied by X. Formally, CF is defined as $CF(X \rightarrow Y) = gain(X \rightarrow Y)/(1 - P_y)$ if $gain(X \rightarrow Y) \geq 0$), and $CF(X \rightarrow Y) = gain(X \rightarrow Y)/P_y$ if $gain(X \rightarrow Y) < 0$).

2.2.2 Relationship Between Quality Measures

The number of objective quality measures is enormous [13], and there is no clear set of metrics appropriate for the pattern mining task. The aforementioned subset of measures (see Table 2.3) comprises different metrics widely used by many experts in the field [13, 27, 29], but we cannot define this group as the best or the optimum one. Considering the previously described group of measures, Fig. 2.4 illustrates how the different pairs of measures are related, which is really interesting to know the behaviour of the metrics. Here, each dot on the scatter plot represents one association rule from different datasets. The position of the dot on the scatterplot represents its A (measure in the x-axis) and B (measure in the y-axis) value. As a matter of example, it is interesting to note that the existing relationship between both support and confidence (see Fig. 2.4) for a varied set of different rules is the same as the one described in Fig. 2.3, which was demonstrated mathematically.

Analysing the properties described by Piatetsky-Shapiro [20], and considering the set of quality measures shown in Table 2.3, the three properties are satisfied by the following metrics: lift, leverage, gain and certainty factor. In this regard, it is interesting to analyse in depth these four quality measures, so the existing relationship between them and the support measure is described below. Despite the fact that support is not considered as a quality measure according to the properties of *Piatetsky-Shapiro*, it is the most well-known metric used in pattern mining [3] and any existing proposal in this field includes this metric. First, we analyse the existing relationship between support ($support(X \rightarrow Y) = P_{xy}$) and lift

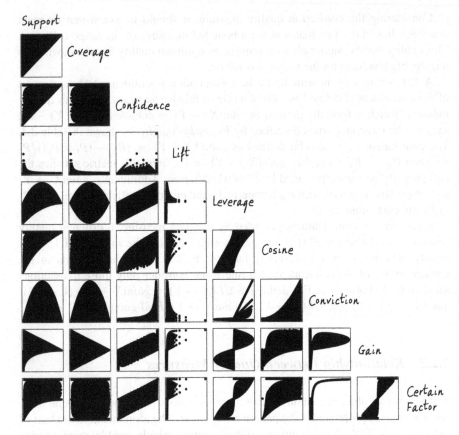

Fig. 2.4 Relationship between pairs of measures

$(lift(X \rightarrow Y) = P_{xy}/(P_x \times P_y))$ as shown Fig. 2.5. As it is illustrated, the smaller the support value the higher the lift value, denoting a high positive dependence between the antecedent X and the consequent Y of an association rule. Nevertheless, if we just analyse the lift values in the range $[0, 1]$, we discover that small support values also imply a high negative dependence between the antecedent X and the consequent Y. Finally, it should be noted that the higher the support the more independent are both X and Y, and a support value of 1 implies that $P_{xy} = P_x \times P_y$ so X and Y are statistically independent.

In a second analysis, we study the relationship between support and leverage, which is defined as $leverage(X \rightarrow Y) = P_{xy} - (P_x \times P_y)$ in the range $[-0.25, 0.25]$. It is quite interesting to note that, similarly to the lift quality measure, leverage denotes independence between X and Y for maximum support values (see Fig. 2.6). For negative leverage values, the behaviour is quite similar to the one obtained for lift. The main different lies on positive leverage values, where the maximum value is bounded. Noted that the upper bound in the lift quality measure is the total number n of transactions in the dataset, whereas in the leverage metric is 0.25 regardless the dataset used.

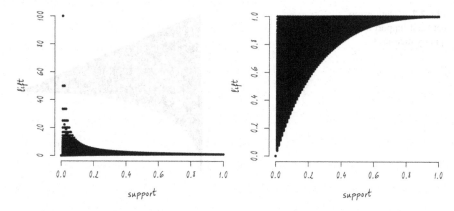

Fig. 2.5 Relationship between support and lift quality measures. Scatter plot on the right illustrates lift values lower than 1, whose distribution is completely different

Fig. 2.6 Relationship between support and leverage quality measures

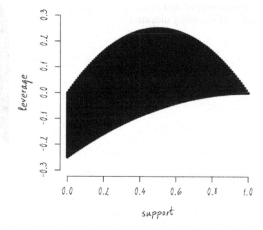

Following with the same analysis and considering now support against gain, which is defined as $gain(X \to Y) = (P_{xy} - (P_x \times P_y))/P_x$, the existing relationship is shown in Fig. 2.7. Similarly to the leverage metric, the gain measure can obtain any value in a bounded range of values, i.e. $[-1, 1]$, which is a great advantage with regard to the lift metric. Furthermore, negative gain values describe a behaviour quite similar to the aforementioned metrics (lift and leverage). In fact, the behaviour described by this metric is quite similar to the one described by the lift. Two are the main differences: (1) the upper bound is delimited, which is a great advantage since it does not depend on the dataset under study and the rules can be properly quantified; (2) the value 0 implies independence between the antecedent X and the consequent Y.

Finally, we analyse the existing relationship between support and certainty factor (CF), which is illustrated in Fig. 2.8. This metric, which is defined as $CF(X \to Y) = ((P_{xy}/P_x) - P_y)/(1 - P_y)$, obtains negative values similarly to the other analysed

Fig. 2.7 Relationship
between support and gain
quality measures

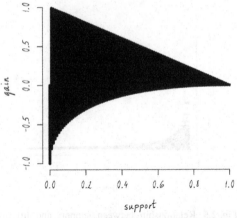

Fig. 2.8 Relationship
between support and certainty
factor (CF) quality measures

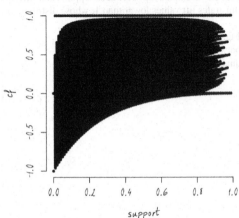

measures, i.e. lift, leverage and gain. Furthermore, a value of 0 for the CF measure
implies an independence between X and Y. Another interesting feature of CF is its
bounded range of values, which is the same as the one of gain, i.e. $[-1, 1]$. The main
difference between CF and the other metrics lies on the fact that high support values
can imply high CF values. Finally, it should be noted that maximum support values,
i.e. $support(X \rightarrow Y) = 1$, produce a value $CF(X \rightarrow Y) = 0$, whereas maximum
CF values can be obtained for any support value lower than 1 and greater than 0.

2.2.3   Other Quality Properties

Since the proposal described by Piatetsky-Shapiro [20], which determined three
different properties to quantify the interest of any metric, many authors have studied
other properties in this regard [27]. The first property (O1) is related to the symmetry

Table 2.4 Summary of properties satisfied by a set of different objective quality measures

Measure	Property 1	Property 2	Property 3	O1	O2	O3	O4
Support	No	Yes	No	Yes	No	No	No
Coverage	No	No	No	No	No	No	No
Confidence	No	Yes	No	No	No	No	Yes
Lift	Yes[a]	Yes	Yes	Yes	No	No	No
Leverage	Yes	Yes	Yes	Yes	Yes	Yes	No
Cosine	No	Yes	Yes	Yes	No	No	Yes
Conviction	No	Yes	No	No	No	Yes	No
Gain	Yes	Yes	Yes	No	No	No	No
Certainty factor	Yes	Yes	Yes	No	No	Yes	No

[a]This property is satisfied just in case that the measure will be normalized

Table 2.5 Sample relative frequencies for an association rule of the form $X \rightarrow Y$

	Y	$\overline{Y}$	Σ
X	0.55	0.05	0.60
$\overline{X}$	0.15	0.25	0.40
Σ	0.70	0.30	1.00

under variable permutation. In this sense, a measure $\mathscr{M}$ satisfies this property if and only if $\mathscr{M}(X \rightarrow Y) = \mathscr{M}(Y \rightarrow X)$. Otherwise, $\mathscr{M}$ is known as an asymmetric measure. As shown in Table 2.4, the symmetric measures considered in this study include support, lift, leverage and cosine. As for asymmetric measures, the following metrics are considered: coverage, confidence, conviction, gain and certainty factor. This type of measures are used for situations in which there is a need to distinguish between the rule $X \rightarrow Y$ and the rule $Y \rightarrow X$.

Let us consider the confidence quality measure, which was defined as P_{xy}/P_x, and the relative frequencies shown in Table 2.5. The confidence quality measure is computed as $confidence(X \rightarrow Y) = 0.55/0.60 = 0.9166$, so this quality measure does not satisfy the property O1, since $confidence(Y \rightarrow X) = 0.55/0.70 = 0.7857$ and, therefore, $confidence(X \rightarrow Y) \neq confidence(Y \rightarrow X)$. On the contrary, the lift quality measure $(P_{xy}/(P_x \times P_y))$ is considered as a symmetric measure since $lift(X \rightarrow Y) = lift(Y \rightarrow X)$. Considering again the relative frequencies shown in Table 2.5, it is obtained that $lift(X \rightarrow Y) = 0.55/(0.60 \times 0.70) = 1.3095 = lift(Y \rightarrow X)$.

The second property (O2) describes the antisymmetry under row/column permutation. A normalized measure $\mathscr{M}$ is antisymmetric under the row permutation operation if $\mathscr{M}(T')=-\mathscr{M}(T)$, considering T as the table of frequencies (see Table 2.2) and T' as the table of frequencies with a permutation on the rows; whereas the measure $\mathscr{M}$ is antisymmetric under the column permutation operation if $\mathscr{M}(T'') = -\mathscr{M}(T)$, considering T'' as the table of frequencies with a permutation on the columns. From the set of metrics used in this study, only the leverage or novelty satisfies the antisymmetry property under row/column permutation (see Table 2.4). According to the authors [27] measures that are symmetric under the row and column permutation

Table 2.6 Row permutation
of the sample relative
frequencies shown in
Table 2.5

	Y	$\bar{Y}$	Σ
X	0.15	0.25	0.40
$\bar{X}$	0.55	0.05	0.60
Σ	0.70	0.30	1.00

Table 2.7 Row and column
permutations of the sample
relative frequencies shown in
Table 2.5

	Y	$\bar{Y}$	Σ
X	0.25	0.15	0.40
$\bar{X}$	0.05	0.55	0.60
Σ	0.30	0.70	1.00

operations do not distinguish well between positive and negative correlations so it should be careful when using them to evaluate the interestingness of a pattern.

Considering the leverage quality measure, which was defined as $P_{xy} - (P_x \times P_y)$, and the relative frequencies shown in Table 2.5, it is possible to assert that this quality measure is antisymmetric under the row permutation (see Table 2.6). Noted that $leverage(X \rightarrow Y) = 0.55 - (0.60 \times 0.70) = 0.13$ on the original contingency table (Table 2.5), and $leverage(X \rightarrow Y) = 0.15 - (0.40 \times 0.70) = -0.13$ when using the row permutation (Table 2.6). Hence, it is possible to assert that leverage satisfies the second property under the row permutation.

A third property (O3) is related to inversion as a special case of the row/column permutation where both rows and columns are swapped simultaneously. This third property describes symmetric binary measures, which are invariant under the inversion operation. A measure $\mathcal{M}$ is a symmetric binary measure if $\mathcal{M}(T) = \mathcal{M}(T''')$, considering T as the table of frequencies (see Table 2.5) and T''' as the table of frequencies with a permutation on both rows and columns (see Table 2.7).

Taking again leverage as a quality measure that satisfies the third property (O3), it is possible to calculate $leverage(X \rightarrow Y) = 0.55 - (0.60 \times 0.70) = 0.13$ on the original contingency table (Table 2.5). On the contrary, when using the table of frequencies with a permutation on both rows and columns (see Table 2.7), the leverage value is calculated as $leverage(X \rightarrow Y) = 0.25 - (0.40 \times 0.30) = 0.13$, so it is demonstrated that the O3 property is satisfied. Finally, let us consider now the confidence measure as a quality measure that does not satisfy this property. Taking the original contingency table (Table 2.5), the confidence value is obtained as $confidence(X \rightarrow Y) = 0.55/0.60 = 0.9166$, whereas for the permutated table (see Table 2.7), the value obtained is $confidence(X \rightarrow Y) = 0.25/0.40 = 0.6250$, so this property is not satisfied by the confidence quality measure.

To complete this analysis, the null-invariant property (O4) is also included, which is satisfied by those measures that do not vary when adding more records that do not contain the two variables X and Y. This property may be of interesting in domains where co-presence of items is more important than co-absence. Table 2.8 shows the relative frequencies when more records that do not contain the two variables X and Y have been added to the original relative frequencies (see Table 2.5). Here, it is possible to demonstrate that confidence satisfies this property since

	Y	$\overline{Y}$	Σ
X	0.275	0.025	0.300
$\overline{X}$	0.075	0.625	0.700
Σ	0.350	0.650	1.000

Table 2.8 Sample relative frequencies when adding more records that do not contain the two variables X and Y to relative frequencies shown in Table 2.5

$confidence(X \rightarrow Y) = 0.55/0.60 = 0.9166$ on the original table of frequencies and this value remains the same for the new table of frequencies, i.e. $confidence(X \rightarrow Y) = 0.275/0.300 = 0.9166$.

2.3 Subjective Interestingness Measures

In previous section, we have described a set of quality measures that provide statistical knowledge about the data distribution. Many of these measures state for the interest and novelty of the knowledge discovered, but none of them describe the quality based on the background knowledge of the user [12]. Sometimes, this background knowledge is highly related to the application domain, and the use of objective measures does not provide useful knowledge.

Many authors have considered that the use of any subjective measure is appropriate only if one of the following conditions are satisfied [11]. First, the background knowledge of users varies due to the application field is highly prone to changes. Second, the interest of the users vary, so a pattern that can be of high interest in a quantum of time t_1 may be useless in a quantum of time t_2. Thus, subjective quality measures cannot be mathematically formulated as objective interestingness measures do.

Knowledge comprehensibility can be described as a subjective measure, in the sense that any pattern can be little comprehensible for a specific user and, at the same time, very comprehensible for a different one [10]. Nevertheless, many authors have considered this metric as an objective measure with a fixed formula: the fewer the number of items, the more comprehensible a pattern is. This concept of comprehensibility can also be applied to the number of patterns discovered [17]. It should be noted that, when an expert want to extract knowledge from a database, it is expected that the knowledge is comprehensible so to provide the user with tons of interesting patterns may be counter-productive. Thus, the fewer the number of patterns provided to the user, the more comprehensible the extracted knowledge is. Nevertheless, the comprehensibility problem is not an easy issue, and it is not only related to the length of the patterns or the number of patterns discovered in the mining process. Comprehensibility can be associated with subjective human preferences and the level of abstraction. As a matter of example, let us consider the representation of the invention of the first light bulb by Thomas Edison. At a low level of abstraction, it is possible to assert that Edison filed his first patent application on October 14th, 1878. Depending on the application domain, this date may be extremely accurate, so for some users it might be enough by describing that the invention of the first light bulb is dated in 1878. Even more, for a young students, it is perfectly fine to know that this invention was carried out in the nineteenth century. The three aforementioned representations are perfectly accurate, but depending on the human preferences, the last representation is more comprehensible than the first one.

The term unexpectedness has also been used in describing subjective quality measures [19]. A pattern is defined as unexpected and of high interest if it contradicts the user's expectations. Let us consider a dataset comprising information about students of a specific course, where the instructor previously knows that 80 % of the students passed the first test. It is expected that a huge percentage of the students also passed the second test according to the previous knowledge. Nevertheless, it is discovered that only 5 % of the students passed this second test. This pattern is completely unexpected to the instructor, denoting that something abnormal has occurred during the second lesson.

In subjective quality measures, actionability plays an important role [25]. A pattern is interesting if the user can do something with it or react to ti to his advantage. Considering the same example described before, which described that only 5 % of the students passed the second test whereas 80 % of these students passed the first test, it is possible to consider this pattern as actionable if it serves to improve both the teaching skills and the resources provided to students. This is a clear example of a pattern that is unexpected and actionable at the same time. Nevertheless, these two concepts are not always associated since an unexpected pattern might not be actionable if the results obtained do not depend on the user but on external systems.

Interactive data exploration is another subjective way of evaluating the quality of the extracted knowledge [28]. This novel task defines methods that allow the user to be directly involved in the discovery process. In interactive data exploration, patterns result to be much more relevant and interesting to the user [6] since he or she explores the data and identifies interesting structure through the visualization and interaction of different models. Nevertheless, this task has an enormous drawback since it can only be applied to relatively small datasets since the mining process cannot be tedious for the end user.

References

1. C. C. Aggarwal and J. Han. *Frequent Pattern Mining*. Springer International Publishing, 2014.
2. R. Agrawal and R. Srikant. Fast Algorithms for Mining Association Rules in Large Databases. In *Proceedings of the 20th International Conference on Very Large Data Bases*, VLDB '94, pages 487–499, San Francisco, CA, USA, 1994. Morgan Kaufmann Publishers Inc.
3. R. Agrawal, T. Imielinski, and A. N. Swami. Mining association rules between sets of items in large databases. In *Proceedings of the 1993 ACM SIGMOD International Conference on Management of Data*, SIGMOD Conference '93, pages 207–216, Washington, DC, USA, 1993.

4. J. L. Balcázar and F. Dogbey. Evaluation of association rule quality measures through feature extraction. In *Proceedings of the 12th International Symposium on Advances in Intelligent Data Analysis*, IDA 2013, pages 68–79, London, UK, October 2013.
5. F. Berzal, I. Blanco, D. Sánchez, and M. A. Vila. Measuring the Accuracy and Interest of Association Rules: A new Framework. *Intelligent Data Analysis*, 6(3):221–235, 2002.
6. M. Bhuiyan, S. Mukhopadhyay, and M. A. Hasan. Interactive pattern mining on hidden data: A sampling-based solution. In *Proceedings of the 21st ACM International Conference on Information and Knowledge Management*, CIKM '12, pages 95–104, New York, NY, USA, 2012. ACM.
7. M. Böttcher, G. Ruß, D. Nauck, and R. Kruse. From Change Mining to Relevance Feedback: A Unified View on Assessing Rule Interestingness Post-Mining of Association Rules: Techniques for Effective Knowledge Extraction. In Y. Zhao, L. Cao, and C. Zhang, editors, *Information Science Reference*, pages 12–37. IGI Global, Hershey, New York, 2009.
8. S. Brin, R. Motwani, J. D. Ullman, and S. Tsur. Dynamic Itemset Counting and Implication Rules for Market Basket Data. In *Proceedings of the 1997 ACM SIGMOD International Conference on Management of Data*, SIGMOD '97, pages 255–264, Tucson, Arizona, USA, 1997. ACM.
9. U. M. Fayyad, G. Piatetsky-Shapiro, P. Smyth, and R. Uthurusamy, editors. *Advances in Knowledge Discovery and Data Mining*. American Association for Artificial Intelligence, Menlo Park, CA, USA, 1996.
10. A. A. Freitas. *Data Mining and Knowledge Discovery with Evolutionary Algorithms*. Springer-Verlag Berlin Heidelberg, 2002.
11. L. Geng and H. J. Hamilton. Interestingness Measures for Data Mining: A Survey. *ACM Computing Surveys*, 38, 2006.
12. B. Goethals, S. Moens, and J. Vreeken. MIME: A Framework for Interactive Visual Pattern Mining. In D. Gunopulos, T. Hofmann, D. Malerba, and M. Vazirgiannis, editors, *Machine Learning and Knowledge Discovery in Databases*, volume 6913 of *Lecture Notes in Computer Science*, pages 634–637. Springer Berlin Heidelberg, 2011.
13. F. Guillet and H. Hamilton. *Quality Measures in Data Mining*. Springer Berlin / Heidelberg, 2007.
14. J. Han and M. Kamber. *Data Mining: Concepts and Techniques*. Morgan Kaufmann, 2000.
15. A. Jiménez, F. Berzal, and J. C. Cubero. Interestingness measures for association rules within groups. In *Proceedings of the 13th International Conference on Information Processing and Management of Uncertainty in Knowledge-Based Systems*, IPMU 2010, pages 298–307. Springer, 2010.
16. N. Lavrač, P. A. Flach, and B. Zupan. Rule Evaluation Measures: A Unifying View. In *Proceedings of the 9th International Workshop on Inductive Logic Programming*, ILP '99, pages 174–185, London, UK, 1999. Springer-Verlag.
17. J. M. Luna, J. R. Romero, C. Romero, and S. Ventura. On the use of genetic programming for mining comprehensible rules in subgroup discovery. *IEEE Transactions on Cybernetics*, 44(12):2329–2341, 2014.
18. A. Merceron and K. Yacef. Interestingness measures for association rules in educational data. In *Proceedings of the 1st International Conference on Educational Data Mining*, EDM 2008, pages 57–66, Montreal, Canada, 2008.
19. B. Padmanabhan and A. Tuzhilin. Unexpectedness as a Measure of Interestingness in Knowledge Discovery. In *Proceedings of the fifth ACM SIGKDD International Conference on Knowledge Discovery and Data Mining*, KDD '99, pages 275–281, New York, NY, USA, 1999. AAAI Press.
20. G. Piatetsky-Shapiro. Discovery, analysis and presentation of strong rules. In G. Piatetsky-Shapiro and W. Frawley, editors, *Knowledge Discovery in Databases*, pages 229–248. AAAI Press, 1991.
21. C. Romero, J. M. Luna, J. R. Romero, and S. Ventura. RM-Tool: A framework for discovering and evaluating association rules. *Advances in Engineering Software*, 42(8):566–576, 2011.
22. D. Sánchez, J. M. Serrano, L. Cerda, and M. A. Vila. Association Rules Applied to Credit Card Fraud Detection. *Expert systems with applications*, (36):3630–3640, 2008.

23. T. Scheffer.   Finding association rules that trade support optimally against confidence. In *Proceedings of the 5th European Conference of Principles and Practice of Knowledge Discovery in Databases*, PKDD 2001, pages 424–435, Freiburg, Germany, 2001.
24. E. H. Shortliffe and B. G. Buchanan. A model of inexact reasoning in medicine. *Mathematical biosciences*, 23:351–379, 1975.
25. A. Silberschatz and A. Tuzhilin. On subjective measures of interestingness in knowledge discovery. In *Proceedings of the 1st international conference on Knowledge Discovery and Data mining*, pages 275–281, Montreal, Quebec, Canada, 1995. AAAI Press.
26. P. Tan and V. Kumar. Interestingness Measures for Association Patterns: A Perspective. In *Proceedings of the Workshop on Postprocessing in Machine Learning and Data Mining*, KDD '00, New York, USA, 2000.
27. P. Tan, V. Kumar, and J. Srivastava. Selecting the right objective measure for association analysis. *Information Systems*, 29(4):293–313, 2004.
28. M. van Leeuwen. Interactive data exploration using pattern mining. In A. H. Gandomi and Conor Alavi, A. H. Ryan, editors, *Interactive Knowledge Discovery and Data Mining in Biomedical Informatics*, volume 8401 of *Lecture Notes in Computer Science*, pages 169–182. Springer Berlin Heidelberg, 2015.
29. C. Zhang and S. Zhang. *Association rule mining: models and algorithms*. Springer Berlin / Heidelberg, 2002.

Chapter 3
Introduction to Evolutionary Computation

Abstract This chapter presents an overview on evolutionary computation, introducing its basic concepts and serving as a starting point for an inexpert user in this field. Then, the chapter discusses paradigms such as genetic algorithms and genetic programming, which are the most widely used techniques in the mining of patterns of interest. Finally, a brief description about other bio-inspired algorithms is considered, paying special interest in ant colony optimization. The main goal is to help the reader to comprehend some basic principles of evolutionary algorithms and swarm intelligence so next chapters of this book can be understood in an appropriate way.

3.1 Introduction

The looking for an optimal search technique has been the aim of many researchers, who first tackled the problem by means of accurate mathematical methods [2]. Nevertheless, the necessity of solving more complex problems gave rise to stochastic methods [20]. Some of these techniques are mainly based on the principles of natural evolution and, more specifically, in the fact that organisms that are capable of acquiring resources will tend to have descendants in the future. It is said that these organisms are more fit for survival, and their characteristics will be selected for their natural descendants.

The process of natural evolution can be modeled algorithmically and simulated by computers, given rise to the concept of evolutionary computation (EC) [14]. Many computer scientists have developed algorithms inspired by natural evolution [24] with the aim of solving problems that are too difficult to tackle with other analytical methods. These artificial evolution algorithms, also known as evolutionary algorithms (EAs), make use of a fitness function that determines how well the solutions solve the predefined problem. These organisms, commonly known as individuals, represent candidate solutions to a predefined problem, and the concept of fitness is quite different to the one of natural evolution, where the fitness of an individual is defined by its reproductive success.

Existing techniques in the EC field enable to abstract the natural evolution process into algorithms used for searching optimal solutions to a specific problem. Since the first use of basic approaches in computer science that copy evolutionary

© Springer International Publishing Switzerland 2016
S. Ventura, J.M. Luna, *Pattern Mining with Evolutionary Algorithms*,
DOI 10.1007/978-3-319-33858-3_3

mechanisms, different kinds of algorithms have been discussed in the literature [1]. EAs are considered as very flexible search methods and can be used to solve numerous problems by considering both a good individual (candidate solution) representation and an accurate fitness function that describes the individual suitability for the problem. Nevertheless, many researchers have criticized the concept of artificial evolution due to its elements of randomness and lack of formal proof of convergence [15].

Focusing on the characteristics of all the EAs, it should be noted that they have some basic elements that are common. The first of these basic elements is the population, which is defined as a pool of two or more candidate solutions or individuals, since it is not possible to deal with the concept of evolution by considering just a single organism [17]. Considering the population of candidate solutions, it is required to maintain a set of individuals that vary from one another to some extend, and this concept is known as diversity. Varied individuals imply the analysis of different areas of the search space, so the concept of diversity is essential in any EA. A second basic element of any EA is the use of mechanisms of inheritance, which are mainly stochastic. These mechanisms will generate new individuals from existing ones, enabling individual characteristics to be transmitted to offspring through generations. Nevertheless, as it is considered in natural evolution, characteristics of the most fit individuals have a higher chance to be transmited to future generations. EA emulates this behavior by means of the selection genetic operator.

Selection of individuals that can breed is another basic element of any EA, describing that the ability to be reproduced depends on environmental factors and it is not completely random. As a matter of example, let us consider that giraffes with slightly longer necks can feed on leaves of higher branches when there is no lower branches. Thus, they have a better chance of survival and this characteristic should be propagated through generations giving rise to giraffes with longer and longer necks. This skill is highly related to the fitness function, which determines the quality of the candidate solution represented by an individual. Hence, the better the fitness function, the more often the individual is selected to breed and to pass its characteristics to later generations of individuals [17].

Regarding the mechanisms to generate new individuals, crossover and mutation operators are widely used. The crossover (recombination) operator produces new individuals by combining genetic materials from other individuals (parents). Offspring will be formed by genetic material from the parents so no new characteristics can be discovered. On the contrary, mutation enables a portion of the genetic material of one individual to be randomly changed to obtain a new genetic material. This operator usually contributes to the population diversity, since it may produce new characteristics that were not previously available in the population. Considering the aforementioned example of giraffes, longer necks might have been produced by a mutation, and this deviant characteristic resulted to be favorable so it was propagated through generations. Finally, note that both operators are stochastic so they are usually applied with predefined probabilities.

Considering the set of EC techniques, it is possible to group them into a number of paradigms. Genetic algorithms (GAs) emphasize crossover over mutation, which is commonly applied with a very low probability. In GAs, individuals were classically represented by binary strings, but nowadays there are new different representations such as strings of real values [6]. Evolution strategies (ES) belong to a second paradigm that was developed as a method to solve real-parameter optimization problems [3]. They typically use an individual representation consisting of a real-valued vector. Initial ES proposals used mutation as the principal operator, but nowadays both mutation and recombination are considered. It should be noted that, in any ES proposal, mutation usually modifies individuals according to a normal distribution. Evolutionary programming (EP) is a third paradigm, which was first defined in [16]. In EP, individuals are represented by a real-valued vector and they are evolved by considering the mutation operator (crossover is not used) with a probability that often follows a normal distribution [7]. Finally, genetic programming (GP) can be considered as a GA with a special encoding. GP was proposed by Koza [27] to build computer programs by means of a complex representation language. The original goal of GP was to find an optimized solution from a search space composed of all possible computer programs. Nevertheless, GP is currently used to evolve other types of knowledge, like rule-based systems [23], since it is considered as an evolutionary and very flexible heuristic technique that enables complex pattern representations to be used. GP represents the solutions to the problem by means of trees [5], enabling any kind of function and domain knowledge to be considered. Solutions represented as trees include two kind of symbols: leaves and internal nodes. Leave nodes correspond to variables and constants, whereas internal nodes correspond to operators and functions.

Another general paradigm widely used to deal with optimization problems is swarm intelligence. Though it does not actually belong to the EC field, it shares several commonalities with it, and has some relevance in pattern mining applications. Instead of simulating the evolution of organisms, swarm intelligence focuses on the interaction among several agents and their environment. In literature, a large number of swarm optimization approaches have been discussed [4, 34], and many researchers have presented different taxonomies [21]. Ant colony optimization (ACO) [11] is a metaheuristic used to find approximate solutions to many optimization problems. ACO is inspired by how ants deposit pheromone on the ground in order to mark some favorable paths that should be followed by other members of the colony in searching for food. Particle swarm optimization (PSO) is another metaheuristic in which particles of a population move through the space of solutions, and they are evaluated according to some fitness criterion. PSO takes its inspiration from the behaviour of bird flocking or avoiding predators [26]. Finally, another swarm intelligence example is artificial bee colony (ABC) [25], which simulates the foraging behaviour of honey bees.

Finally, it is interesting to describe how all the aforementioned paradigms are related and which features are shared by subsets of them. In this regard, Fig. 3.1 illustrates a classification of the set of paradigms described in this section.

Fig. 3.1 Relationships
between the set of paradigms
analised in this chapter

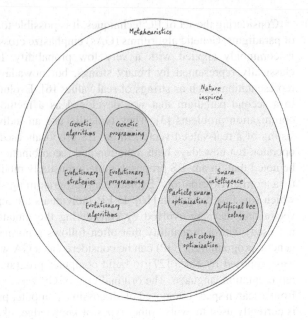

3.2 Genetic Algorithms

Genetic algorithms (GAs) are considered as one of the pillars in which the
evolutionary algorithms (EAs) rest [1]. As any EA, a set of individuals that
represents candidate solutions to the problem under study is considered by GAs,
and these individuals are often represented by means of a fixed-length string or
vector of values. Each of these values is known under the term *gene*, representing a
characteristic or a small part of the whole candidate solution. In following sections,
the standard procedure of any GA and the individual representation are discussed,
describing how the use of a fixed-length string of values enables a candidate solution
to be represented. Additionally, different mechanisms for both generating new
individuals and selecting optimal individuals are described.

3.2.1 Standard Procedure

A GA is an iterative procedure that operates on a population [18], which was
previously defined as a pool of two or more individuals (candidate solutions). In
this iterative procedure (see Fig. 3.2), the GA first generates the initial population
by following a heuristic or, in many cases, by a random way. Subsequently, the
iterative procedure is run, where each iteration is known as *generation*. Any new

Require: *population_size*   {Number of individuals to be considered in the population}
Ensure: *population*
 1: *population* ← ∅
 2: *parents* ← ∅
 3: *offspring* ← ∅
 4: *population_size* ← *n*
 5: *population* ← generate(*population_size*)
 6: **for** ∀*member* ∈ *population* **do**
 7:    evaluate(*member*)
 8: **end for**
 9: **while** termination condition is not reached **do**
10:    *parents* ← select(*population*)
11:    *offspring* ← apply genetic operators on the set *parents*
12:    **for** ∀*member* ∈ *offspring* **do**
13:       evaluate(*member*)
14:    **end for**
15:    *population* ← update(*population*,*offspring*)
16: **end while**

Fig. 3.2 Standard GA procedure

individual generated either by the initial population or by the genetic operators is evaluated and assigned a fitness value that determines how promising the individual is and how *close* it is from the optimal solution.

In each generation of the evolutionary schema, a subset of individuals from the population is selected to act as parents according to some heuristic. Most GAs use a selection procedure based on the fitness function, so the better the fitness of an individual the more probable this individual is to be selected as a parent. This set of parents is used to form a new population by means of genetic operators so the most fit individuals are used to breed and to pass their characteristics to later generations.

Two are the operators used to produce new individuals in GAs [33], which are widely known as crossover and mutation. Crossover is considered as the principal genetic operator, and it usually takes two individuals from the set of parents and produces one (or two) new individual by combining characteristics of the parents. As for the mutation operator, it is usually considered as a secondary operator and it is typically used with a very low probability. This secondary operator applies a perturbation on the genes of a single individual, usually a subset of them. The aim of this genetic operator is to help in preventing premature convergence since the new substring might comprises unexplored characteristics.

Finally, it is interesting to note that any GA is considered as a stochastic iterative procedure that cannot guarantee that the optimal (global) solution is found [1]. This iterative procedure is run while the termination condition is not reached, which can be determined by a specific number of generations (iterations) or number of invocations of the objective function, or by a more specific procedure according to the solutions found.

3.2.2 Individual Representation

As mentioned above, GAs typically consider a set of individuals that represents candidate solutions to the problem under study, and these individuals are often represented by means of a fixed-length string or vector of values. In the field of GAs, individuals were first represented by binary strings [24], so initial genetic operators were also designed for this kind of representation. This is the simplest way of representing an individual, since each gene can take either the value 1 or 0—true and false logical values, respectively.

According to the aforementioned individual representation, many problems have been solved by using it, e.g. the knapsack problem. In the knapsack problem, a set of n distinct objects is considered to be included in a knapsack. Each object x_i might be present or not, and it has either a weight w_i and a different price p_i. It is required that the knapsack does not exceed a specific weight W, i.e. $W \leq \sum_{i=1}^{n} x_i \times w_i$. To achieve the desired solution, it is required to determine whether each object i is included or not in the knapsack—x_i takes the value 1 or 0 to represent the presence or absence of the object in the knapsack, respectively. Thus, it is possible to represent each individual as a binary string where each specific position on the string is associated with a specific object. The aim of this problem is to include objects into the knapsack in order to maximize its profit (according to the profits of each object), i.e. $P = \sum_{i=1}^{n} x_i \times p_i$ should be maximum.

Many authors have supported that the use of binary strings is not always the most accurate representation [18], specially when continuous variables need to be represented. In such situation, the results will depend on the precision (number of bits) to represent the values for the continuous variables. The lost of precision can be reduced by using a larger number of genes (values 0 or 1). In order to solve this problem, some authors have included real values into the individual representation [6]. In general, it is taught [31] that a better performance can be achieved if the individual representation incorporates both explicit knowledge for the problem and specific genetic operators. In this regard, the genetic operators combine different parts of the parents, producing valid new offspring (new solution candidates). Hence, the genetic operators should be accurately designed depending on the individual representation.

3.2.3 Genetic Operators

The suitability of each individual to the specific problem is ranked by using a fitness function, which determines how well the individual solves the predefined problem. As denoted in previous sections, the concept of fitness in natural evolution is related to the reproductive success of an individual so the better the fitness, the more often the individual is selected to breed and to pass its characteristics to later generations [17]. In order to emulate this behaviour, researchers in artificial evolution have

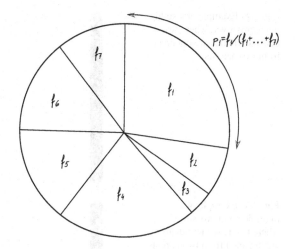

Fig. 3.3 Example of roulette wheel where the size of each sector is proportional to the fitness value, where $\sum_{j=1}^{7} f_j = 1$

proposed numerous methods to select individuals from the population to produce new ones, and the selection probability is proportional to the fitness function.

The most straightforward implementation of this rule is the so-called roulette-wheel selection [22]. This method assumes that the probability of selection of an individual is proportional to its fitness value when all the fitness values of the remaining individuals are considered. Let us consider n individuals where the fitness value of a specific individual i is denoted as f_i. The selection probability of an individual i is given by the probability $p_i = f_i / \sum_{j=1}^{n} f_j$. For a better understanding, let us consider a roulette wheel (see Fig. 3.3) having sectors with different sizes, and these sizes are proportional to the fitness value f_i of each individual i. Thus, selecting a random individual according to their fitness values evokes the process of spinning the wheel. By repeating this process, individuals are chosen using random sampling with replacement.

Many authors have proposed other selection methods [13, 38], the tournament selection being one of the most well-known and many different variants can be found in the literature [32]. From all the existing variants, the most widely used is the k-tournament selection where k individuals are randomly selected from the population and the best one (according to their fitness values) from the k selected ones is considered for reproduction.

Once a set of individuals have been chosen by a selection method, new individuals are obtained (thanks to recombination and mutation) from the ones previously selected. Beginning with the crossover operator, it usually takes two individuals from the set of parents and cut their string of values at some randomly chosen position (one-point crossover operator). The substrings are swapped and new individuals are obtained, resulting in the combination of the characteristics of both parents. This crossover technique is known as single point crossover or one point crossover (see Fig. 3.4), and it is one of the simplest crossover techniques that we can

Fig. 3.4 Example of a single point crossover, where two new individuals are obtained from two parents

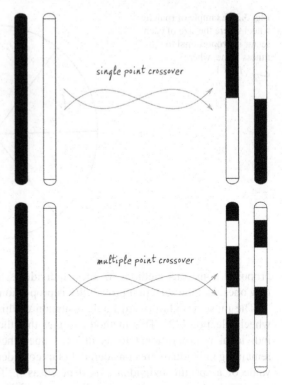

Fig. 3.5 Example of a multiple point crossover, where two new individuals are obtained from two parents

find in the literature. An extension of this technique is the multiple point crossover, which randomly chooses n cut points for each string of values and these substrings are then swapped (see Fig. 3.5).

Finally, another well-known crossover operator is known as uniform crossover, where each gene of the offspring is computed by copying the corresponding gene from one of the parents selected at random. Genes are combined with a predefined probability so values far from 0.5 produces offsprings that are similar to one of the parents. Many researchers [22] have considered the value 0.5 as the standard probability value, but it might vary depending on the problem being solved.

Unlike crossover operators, which work on two parents to obtain offspring, mutation is an operator that acts on a single individual at a time. Though there are several mutation proposals, according to the tackled problem and the representation, the most classic one replaces the value of a specific gene with a new value randomly generated. Both the gene to be mutated and the new value for this gene are randomly chosen, so a major feature of this operator is its ability to introduce a gene value that was not present in the population, increasing the genetic diversity. In general, mutation is usually applied with a small probability in GAs.

3.3 Genetic Programming

Genetic programming (GP) is an evolutionary and very flexible heuristic technique that has been considered by many authors as an extension of GAs [35]. GP represents the solutions to the problem by means of trees [8], enabling any kind of function and domain knowledge to be considered. Solutions represented as trees include two kind of symbols: leaves and internal nodes. Leave nodes correspond to variables and constants, whereas internal nodes correspond to operators and functions. There are GP algorithms that use other kind of individual representations [5], some of them are able to manipulate different data types and allow to define the problem constraints [36]. This way of including constraints is named grammar-guided genetic programming (GGGP or G3P) [40], which employs a context-free grammar to generate any feasible solution to the problem under study [30]. Thus, the grammar constrains the search space and solutions are generated by applying a set of productions rules.

GP was proposed as an extension of GAs to build computer programs by means of a complex representation language [23]. The first aim of GP was to find an optimized solution from a search space composed by all possible computer programs. Nowadays, GP is currently used to evolve other types of knowledge, like rule-based systems [39].

In following sections, the individual representation considered by GP is described. Additionally, different mechanisms for generating new individuals are analysed, considering either crossover and mutation operators. Finally, the code bloat problem is introduced and studied in depth.

3.3.1 Individual Representation

In GP, individuals are typically represented as tree structures rather than strings of values. Each tree representation includes two different groups of symbols. First, leave nodes correspond to terminal symbols, which comprise variables and constants. Second, internal nodes describe operators and functions, and they are widely known as non-terminal symbols. In general, individuals in GP usually grow in size and shape unlike conventional GAs that use a fixed-length string of values. Nevertheless, there are some GAs that use a variable-length representation and there are also some GP algorithms that use some limit of the tree size. Thus, it is possible to assert that the main difference between both GA and GP is that the latter contains not only values but also functions.

Figure 3.6 shows a sample GP individual represented as a tree structure. Terminal symbols considered in this individual are given by the constants 2 and 7, and also by the variable x. Additionally, this individual comprises three different internal nodes or operators: product ($*$), sum ($+$) and division ($/$). GP usually represents the expressions denoted by the tree structure by using a prefix notation, which makes

Fig. 3.6 Example of an
individual representation by
means of a tree shape

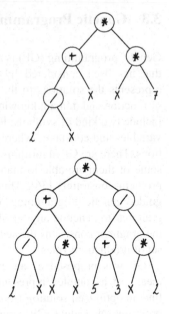

Fig. 3.7 Example of an
individual generated by the
full method so the depth of all
the leaves is the same

easier to obtain the relationship between subexpressions and their corresponding
subtrees. Thus, the GP individual illustrated in Fig. 3.6 represents the expression
$(* (+ (/ 2 X) X) (* X 7))$.

The generation of individuals in GP is a stochastic procedure like in other EAs
found in the literature. Many authors have considered different approaches to obtain
random individuals [35], and two of the first methodologies in this regard are known
as full and grow methods. In both methods a maximum predefined depth is used so
no extremely large individuals can be generated. The depth of a specific node within
the tree is defined as the number of nodes needed to be traversed to reach it (starting
from the root node, whose depth is 0). Similarly, the depth of a tree is defined as
the largest depth that can be found in the tree by considering its leaves or terminal
symbols. Considering the sample individual shown in Fig. 3.6, the depth of the tree
is 3, whereas the depth of the node that represents the sum $(+)$ is 1.

The full method generates trees having leaves which are at the same depth. In this
method, nodes are taken at random from the set of functions until the maximum tree
depth is reached and, at this point, only terminal nodes can be chosen. Notice that
the fact that all the leaves have the same depth does not necessarily imply that all
the generated trees have the same shape. In fact, it only happens if all the functions
have an equal arity, also considered as the number of subtrees. Considering now the
sample individual illustrated in Fig. 3.7, it is possible to state that this is a full tree
since the depth of every leaf node is the same, 3 in this case.

As for the grow method, it enables a great variety of sizes and shapes to be
generated. Unlike full method, the grow method can randomly choose any node
from either the set of functions or the set of terminal values. The only requirement
to be considered in this method is the maximum depth of the tree, so only terminal

symbols can be selected once the maximum depth is reached. Figure 3.6 shows a sample individual that was obtained by the aforementioned grow method.

Finally, it should be noted that these methods often make it difficult to control the sizes and shapes of the generated individuals. For example, both the sizes and shapes of the individuals obtained by means of a grow method are highly sensitive to the number of functions and terminal symbols. If the number of terminal symbols is higher than the number of functions, then the generation procedure is biased to obtain relatively short trees. On the contrary, if the number of terminal symbols is lower than the number of functions, then the grow method tends to behave similarly to the full one.

3.3.2 Genetic Operators

Since GP is defined as an extension of GAs [35], GP also uses the two main genetic operators considered by any GA, which enables new solutions to be produced. Similarly to any GA, crossover is considered as the major operator, taking two individuals to act as parents and producing new offspring by swapping genetic material of the parents. Nevertheless, typical crossover operators used in GAs cannot be used in GP since the representation of individuals is completely different [17].

In any GP algorithm, crossover operators usually take a randomly chosen subtree from a parent and it is swapped for another randomly chosen subtree from another parent. Figure 3.8 shows two sample parents denoted by the expressions

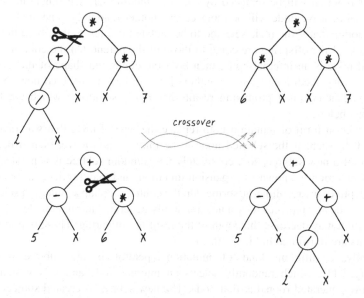

Fig. 3.8 Example of a basic GP crossover to obtain two new individuals

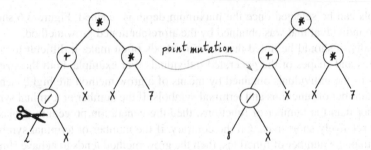

Fig. 3.9 Example of point mutation where the terminal node 2 is replaced by the terminal node 8

$(* (+ (/ 2 X) X) (* X 7))$ and $(+ (- 5 X) (* 6 X))$, which produce two new individuals denoted by the expressions $(* (* 6 X) (* X 7))$ and $(+ (- 5 X) (+ (/ 2 X) X))$. Many studies have evaluated the impact of the crossover operator on the individuals [5]. It is obtained that GP crossover tends to have a certain destructive effect on the individuals, and it might be considered as a kind of driven mutation operator where the genetic material to be inserted into the offspring is not randomly generated but obtained from material that already exists in the population.

Let us consider now the GP mutation, which produces a new individual from a single parent. In GP, the mutation genetic operator is applied with a probability value lower than the one used for crossover. Basic GP mutations randomly choose a subtree of an individual and generates a completely new subtree. The first and simplest form of mutation is known as point mutation (see Fig. 3.9), which replaces a single node in the tree with another node of the same type. Thus, an internal node or function will be replaced by another randomly-chosen function, whereas a terminal or leave node will be replaced by another randomly-generated terminal node. Notice that if the node selected to be mutated is a leave, then both the depth and shape of the offspring are equal to those of the parent. On the contrary, if the selected node is an internal node, then both the depth and shape of the resulting tree may vary. Additionally, some authors [19, 29] have considered new forms of mutation that obtained promising results in finding solutions, which are briefly described below.

A different form of mutation is to replace the type of node that was randomly selected. In such, if the selected node to be mutated is a leave, then it becomes the root of a new subtree, so a completely new random subtree is generated. This type of mutation is known as expansion mutation, and may produce an increment in the depth of the resulting offspring. On the contrary, collapse mutation selects an internal node and replaces it by a new randomly generated terminal node. This type of mutation may decrease the depth of the resulting offspring. These two mutation operators are illustrated in Fig. 3.10.

Finally, a different kind of mutation operator is the subtree mutation (see Fig. 3.11), which randomly selects an internal node and a new subtree is randomly generated rooted at that node. The new subtree is created subject to the depth and/or size restrictions.

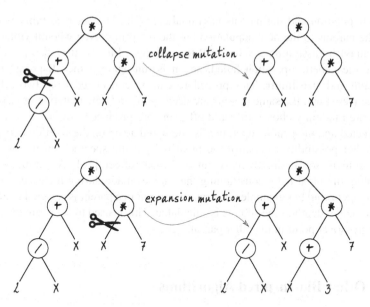

Fig. 3.10 Examples for the collapse and expansion mutations

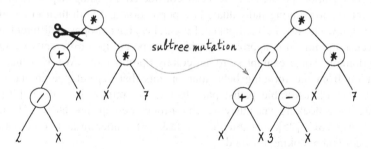

Fig. 3.11 Example of subtree mutation where a completely new subtree is obtained

3.3.3 Code Bloat

Many researchers have demonstrated that the average size (number of nodes) of the GP individuals starts growing after a certain number of generations [5]. Typically this incremental size is not accompanied by any corresponding increase in fitness, and this phenomenon is known as bloat.

To have control of bloat is a crucial issue in GP, and many techniques have been proposed in this sense [28, 37]. The simplest method to control the growth of the trees is to use limits on the size and depth of the offspring obtained by the genetic operators. Thus, it is possible to apply a genetic operator and then check whether the offspring satisfies the depth and size limits. In such a situation where the new generated individual does not satisfies these limits, then it cannot be added to the population and the parents pass unaltered [9]. Another possibility is not to pass the parents to the new population but the offspring (even when it violates the constraints) by assigning a value of 0 to its fitness. Finally, it is also possible to declare the genetic operation failed, and it is run again if the offspring violates the depth and size limits. This procedure can be considered from two different perspectives. First, the same parents are used again and new mutation or crossover points are randomly chosen till valid offspring are produced. Second, new parents are selected and the genetic operation is attempted again on these different parents.

Another possibility to control the bloat is by using specific genetic operators, which directly or indirectly have an anti-bloat effect [10]. A simple way of controling the bloat is by constraining the choices made during the execution of a genetic operation. For example, the crossover point in a parent is selected randomly, and the size or depth of the subtree is calculated. This is used to constrain the choice of the crossover point in the other parent.

3.4 Other Bio-Inspired Algorithms

Many optimization problems have been considered by using a paradigm based on the interaction among individuals in a population and with their environment, which is known as swarm intelligence. Researchers have proposed different swarm intelligence optimization approaches, which can be grouped differently depending on the domain. Some examples are ant colony optimization (ACO)v, which takes inspiration from the foraging behaviour of ants that deposit pheromone on the ground to mark favorable paths; particle swarm optimization (PSO) [26], that simulates birds flocking when they search for food or escape predators; and artificial bee colony (ABC) [25], that imitates the task differentiation and coordination of honey bees when looking for food.

From all the existing algorithms in swarm intelligence, ACO [12] has been considered in the mining of patterns of interest [34]. ACO is considered as an approximate algorithm used to obtain good enough solutions to hard combinatorial optimization problems in a reasonable amount of computation time. This kind of algorithms is inspired in the real behavior of ants when searching for food, which initially explore the area surrounding their nest in a random manner. Once a specific ant finds a resource of food, it evaluates both the quality and the quantity of that resource and takes a portion of food back to the nest. In order to remember the path and to guide other ants to the food, the ant deposits a chemical pheromone trail on the ground, and the quantity of pheromone used depends on both the quantity and quality of the food discovered. In such a way, if an ant has a choice of paths to follow, the preferred route is the one with the highest amount of pheromone.

In ACO algorithms, ants wander through a graph, what represents the decisions it takes at each moment. At the beginning, each ant is situated on one node of the graph, and is associated to an empty solution. Then, the ant stochastically chooses one of the outgoing arcs, according to their weights and pheromone amounts, and this action is associated with a decision that is included in its solution. The ant repeats this process from the new node until the solution is complete. At the same time, the pheromone amounts of the arcs are updated, so ants arriving at the corresponding nodes will evaluate their options differently.

References

1. M. Affenzeller, S. Winkler, S. Wagner, and A. Beham. *Genetic Algorithms and Genetic Programming: Modern Concepts and Practical Applications.* Chapman & Hall/CRC, 1st edition, 2009.
2. G. Allaire. *Numerical analysis and optimization: an introduction to mathematical modelling and numerical simulation.* Numerical Mathematics and Scientific Computation. Oxford University Press, New York, NY, 2007.
3. T. Bäck, F. Hoffmeister, and H. S. Schwefel. A Survey of Evolution Strategies. In *Proceedings of the Fourth International Conference on Genetic Algorithms*, pages 2–9, San Francisco, CA, USA, 1991. Morgan Kaufmann.
4. S. Bansal, D. Gupta, V. K. Panchal, and S. Kumar. Swarm intelligence inspired classifiers in comparison with fuzzy and rough classifiers: a remote sensing approach. In *Proceedings of the 2nd International Conference on Contemporary Computing*, IC3 2009, pages 284–294, Noida, India, 2009.
5. W. Banzhaf, F. D. Francone, R. E. Keller, and P. Nordin. *Genetic programming: an introduction. On the automatic evolution of computer programs and its applications.* Morgan Kaufmann Publishers Inc., San Francisco, CA, USA, 1998.
6. M. Bessaou and P. Siarry. A genetic algorithm with real-value coding to optimize multimodal continuous functions. *Structural and Multidisciplinary Optimization*, 23(1):63–74, 2002.
7. H. G. Beyer and H. P. Schwefel. Evolution strategies - a comprehensive introduction. *Natural Computing: an international journal*, 1(1):3–52, 2002.
8. A. Cano, J. M. Luna, A. Zafra, and S. Ventura. A classification module for genetic programming algorithms in JCLEC. *Journal of Machine Learning Research*, 16:491–494, 2015.
9. E. F. Crane and N. F. McPhee. The effects of size and depth limits on tree based genetic programming. In T. Yu, R. L. Riolo, and B. Worzel, editors, *Genetic Programming Theory and Practice III*, volume 9 of *Genetic Programming*, chapter 15, pages 223–240. Springer, 2005.
10. R. Crawford-Marks. Size control via size fair genetic operators in the pushGP genetic programming system. In *In Proceedings of 2002 the Genetic and Evolutionary Computation Conference*, GECCO 2002, pages 65–79, New York, USA, 2002. Morgan Kaufmann Publishers.
11. M. Dorigo and T. Stützle. *Ant Colony Optimization*. Bradford Company, Scituate, MA, USA, 2004.
12. M. Dorigo and C. Blum. Ant colony optimization theory: A survey. *Theoretical Computer Science*, 344(2–3):243–278, 2005.
13. D. Dumitrescu, B. Lazzerini, L. C. Jain, and A. Dumitrescu. *Evolutionary Computation.* CRC Press, Inc., Boca Raton, FL, USA, 2000.
14. A. E. Eiben and J. E. Smith. *Introduction to Evolutionary Computing.* SpringerVerlag, 2003.
15. D. Floreano and C. Mattiussi. *Bio-Inspired Artificial Intelligence: Theories, Methods, and Technologies.* The MIT Press, 2008.
16. L. J. Fogel, A. J. Owens, and M. J. Walsh. *Artificial intelligence through simulated evolution.* Wiley, Chichester, WS, UK, 1966.
17. A. A. Freitas. *Data Mining and Knowledge Discovery with Evolutionary Algorithms.* Springer-Verlag Berlin Heidelberg, 2002.

18. A. A. Freitas. A Survey of Evolutionary Algorithms for Data Mining and Knowledge Discovery. In A. Ghosh and S. Tsutsui, editors, *Advances in Evolutionary Computing*, pages 819–845. Springer-Verlag New York, Inc., New York, NY, USA, 2003.

19. M. Fuchs. Crossover versus mutation: an empirical and theoretical case study. In *In Proceedings of the third annual conference on Genetic Programming*, GP '98, pages 78–85, 1998.

20. M. Gendreau and J. Potvin. *Handbook of Metaheuristics*. Springer Publishing Company, Incorporated, 2nd edition, 2010.

21. L. Goel, D. Gupta, V. K. Panchal, and A. Abraham. Taxonomy of nature inspired computational intelligence: a remote sensing perspective. In *Proceedings of the 4th World Congress on Nature and Biologically Inspired Computing*, NaBIC 2012, pages 200–206, Mexico City, Mexico, 2012.

22. D. E. Goldberg. *Genetic Algorithms in Search, Optimization and Machine Learning*. Addison-Wesley Longman Publishing Co., Inc., Boston, MA, USA, 1st edition, 1989.

23. P. González-Espejo, S. Ventura, and F. Herrera. A Survey on the Application of Genetic Programming to Classification. *IEEE Transactions on Systems, Man and Cybernetics: Part C*, 40(2):121–144, 2010.

24. J. H. Holland. *Adaptation in Natural and Artificial Systems*. The University of Michigan Press, 1975.

25. D. Karaboga, B. Akay, and C. Ozturk. Artificial Bee Colony (ABC) Optimization Algorithm for Training Feed-Forward Neural Networks. In *Proceedings of the 4th International Conference on Modeling Decisions for Artificial Intelligence*, MDAI '07, pages 318–329, Kitakyushu, Japan, 2007.

26. J. Kennedy and R. C. Eberhart. Particle swarm optimization. In *Proceedings of the 1995 IEEE International Conference on Neural Networks*, volume 4, pages 1942–1948, 1995.

27. J. R. Koza. *Genetic Programming: On the Programming of Computers by Means of Natural Selection (Complex Adaptive Systems)*. A Bradford Book, 1 edition, 1992.

28. W. B. Langdon, T. Soule, R. Poli, and J. A. Foster. The evolution of size and shape. In L. Spector, W. B. Langdon, U. O'Reilly, and P. J. Angeline, editors, *Advances in Genetic Programming III*, chapter 8, pages 163–190. MIT Press, Cambridge, MA, USA, June 1999.

29. S. Luke and L. Spector. A comparison of crossover and mutation in genetic programming. In *Proceedings of the Second Annual Conference on Genetic Programming*, GP '97, pages 240–248. Morgan Kaufmann, 1997.

30. R. McKay, N. Hoai, P. Whigham, Y. Shan, and M. O'Neill. Grammar-based Genetic Programming: a Survey. *Genetic Programming and Evolvable Machines*, 11:365–396, 2010.

31. Z. Michalewicz. *Genetic Algorithms + Data Structures = Evolution Programs*. Springer-Verlag, London, UK, UK, 1996.

32. B. L. Miller and D. E. Goldberg. Genetic algorithms, selection schemes, and the varying effects of noise. *Evolutionary Computation*, 4:113–131, 1996.

33. M. Mitchell. *An Introduction to Genetic Algorithms*. MIT Press, Cambridge, MA, USA, 1998.

34. J. L. Olmo, J. M. Luna, J. R. Romero, and S. Ventura. Mining association rules with single and multi-objective grammar guided ant programming. *Integrated Computer-Aided Engineering*, 20(3):217–234, 2013.

35. R. Poli, W. B. Langdon, and N. F. McPhee. *A Field Guide to Genetic Programming*. Lulu Enterprises, UK Ltd, 2008.

36. A. Ratle and M. Sebag. Genetic Programming and Domain Knowledge: Beyond the Limitations of Grammar-Guided Machine Discovery. In *Proceedings of the 6th International Conference on Parallel Problem Solving from Nature*, PPSN VI, pages 211–220, Paris, France, September 2000.

37. T. Soule and J. A. Foster. Effects of code growth and parsimony pressure on populations in genetic programming. *Evolutionary Computation*, 6(4):293–309, 1998.

38. D. Thierens and D. Goldberg. Convergence models of genetic algorithm selection schemes. In Y. Davidor, H. P. Schwefel, and R. Männer, editors, *Parallel Problem Solving from Nature - PPSN III*, volume 866 of *Lecture Notes in Computer Science*, pages 119–129. Springer Berlin Heidelberg, 1994.

39. A. Tsakonas, G. Dounias, J. Jantzen, H. Axer, B. Bjerregaard, and D. G. von Keyserlingk. Evolving rule-based systems in two medical domains using genetic programming. *Artificial Intelligence in Medicine*, 32(3):195–216, 2004.
40. M. L. Wong and K. S. Leung. *Data Mining Using Grammar-Based Genetic Programming and Applications*. Kluwer Academic Publishers, Norwell, MA, USA, 2000.

Chapter 4
Pattern Mining with Genetic Algorithms

Abstract This chapter describes the use of genetic algorithms for the mining of patterns of interest and the extraction of accurate relationships between them. The current chapter first makes an analysis of the utility of genetic algorithms in the mining of patterns of interest, paying special attention to the computational time and the memory requirements. Then, it describes general issues for any genetic algorithm in the pattern mining field, explaining different ways of representing patterns such as the one used to extract continuous patterns, which include richer information. Additionally, different genetic operators and fitness functions are properly described, denoting their usefulness in the mining of both patterns of interest and accurate associations. Then, different algorithmic approaches in the pattern mining field are analysed. Finally, this chapter deals with a series of application domains in which genetic algorithms for mining either patterns and association rules have been successfully applied.

4.1 Introduction

Pattern mining [13] and the discovery of relationships between patterns [41] are considered as really interesting tasks for the extraction of comprehensible, useful, and non-trivial knowledge from large databases. In its origins, the pattern mining task was strongly associated with the market basket analysis, enabling the analysis of specific products that are strongly related and which tend to be bought on impulse [6]. With the growing interest in the data storage, more and more information needs to be analysed, so the use of traditional approaches for mining patterns of interest may suffer from both computational time and memory requirements [34]. According to some authors [2], pattern mining is a task of high interest due to the computational challenge issues. The pattern mining task turns into an arduous task depending on the search space, which exponentially grows with the number of single items or singletons considered into the application domain. For a better understanding, let us consider a set of items $I = \{i_1, i_2, \ldots, i_n\}$ in a dataset. A maximum number of $2^{|I|} - 1$ different patterns can be found in a dataset comprising $|I| = n$ singletons, so a straightforward approach becomes extremely complex with the increasing number of singletons. Besides, data in real-world

© Springer International Publishing Switzerland 2016
S. Ventura, J.M. Luna, *Pattern Mining with Evolutionary Algorithms*,
DOI 10.1007/978-3-319-33858-3_4

applications usually consist of continuous values so traditional approaches in the pattern mining field require a splitting of the continuous values to transform them into discrete values.

Genetic algorithms (GAs) have emerged as robust and practical optimization methods [12] that can be applied either on discrete and continuous domains. According to the characteristics described by any GA [17], some basic elements are common. The first of these basic elements is the concept of individual, which is defined as a string of values that represents a feasible solution to the problem under study [7]. Each of these values can be defined as genes, which describe important characteristics of the individual represented. For many authors, the string of genes is also known as chromosome [11], and the length and structure of the chromosome varies depending on the information to be represented. A second basic element in any GA is the concept of population, which is defined as a pool of two or more individuals. Considering the population as a set of individuals or candidate solutions, the diversity among these solutions is an essential issue in any evolutionary algorithm (EA). Thus, individuals from the population vary from one another to some extend, which implies the analysis of different areas of the search space. An additional basic element of any GA is the use of mechanisms that enable new individuals to be randomly generated from existing ones. One of the main features of these mechanisms is their ability to transmit parts of the individual characters to offspring, so important features of any individual will not be lost through generations.

The use of GAs for mining patterns of interest and learning association rules is more and more common [8]. Thus, the pattern mining problem can be considered as a combinatorial optimization problem. In general terms, the schema followed by any GA in the pattern mining field is as follows: (1) the initial population is formed by randomly generating individuals that represent patterns or rules according to some prerequisites; (2) all the individuals generated in the initial population are evaluated by means of a fitness function, which is usually based on some of the objective and subjective quality measures described in Chap. 2; (3) a subset of individuals from the population are selected to act as parents; (4) a series of genetic operators (crossover and mutation) are applied on the subset of parents in order to produce new individuals; (5) all the new individuals generated are evaluated according to a fitness function; (6) all or some individuals from the current population are replaced by the new individuals generated with the genetic operators. Finally, it should be noted that steps (3)–(6) are repeated a predefined number of times, each repetition being defined as a generation of the evolutionary process.

All the aforementioned steps need to be properly designed to reach to an optimal solution or a set of optimal solutions. Different authors [4] have proposed interesting methodologies to generate the initial population, paying special attention to the way in which the diversity is maintained. The individual encoding has also been studied by many researchers [25, 40], and many different proposals have been described. GAs have been used to represent patterns in a binary way, where each gene describes whether a specific item appears or not in the database. Continuous patterns have also been represented by means of GAs, not requiring a previous splitting of the continuous values to transform them into discrete values. Additionally, the use of negative patterns have been tackled by GAs, describing a negative relationship between items. Finally, GAs have been used not only for mining patterns of interest [26] but also association rules [4] that describe a relationship between two disjoint sets of items.

According to the process of generating new individuals from a set of parents, traditional operators [27] have been considered by many authors in the mining of patterns of interest. However, not all the genetic operators can be useful for all the encoding criterion, so different proposals are required depending on the way in which individuals are encoded [40]. Finally, it should be noted that the way in which every individual is evaluated is crucial. Thus, the set of new individuals produced by the genetic operators is required to be analysed to determine how close each individual is from the solution. In this regard, different fitness functions have been proposed, most of them considering some of the metrics described in Chap. 2, which were divided into subjective and objective quality measures.

Finally, as it was previously described, the market basket analysis was the first application domain in which pattern mining and association rules were applied. Since its definition in the early 1990s [3], the pattern mining task has provoked tremendous curiosity among experts on different application fields [31, 33]. In many of these fields, authors have considered the use of GAs to solve the problem under study. For instance, medical domains have been analysed by means of a GA that extracts interesting relationships between attributes [18]. Additionally, GAs have also been used for the discovery of association rules from the total ozone mapping spectrometer on board the NASA Nimbus-7 satellite, which measures three different locations of the Iberian Peninsula [24]. Another different field in which the mining of association rules by means of GAs has been applied to is in microarray analysis [23], describing interrelations between genes.

4.2 General Issues

As previously described in Chap. 3, GAs are considered as one of the pillars in which EAs rest [1]. Any GA includes a set of individuals that represents candidate solutions to the problem under study. Focusing on the pattern mining problem, many authors have considered the use of GAs as a way to discover patterns of high interest for the end user. In the following sections, different individual representations are discussed, describing how the use of different strings of values enables any kind of patterns to be represented. Furthermore, different mechanisms for generating new individuals that represent patterns are described. Finally, we analyse different ways of quantifying the interest of the generated individuals.

4.2.1 Pattern Encoding

One of the easiest ways to represent a set of items in GAs is the use of a string of binary values [35]. Following this encoding, each gene or single value within the string accounts for a given item [36] so given the set of items $I = \{i_1, i_2, \ldots, i_{10}\}$ in a dataset, then the string 1001100101 matches the pattern $\{i_1, i_4, i_5, i_8, i_{10}\}$. Nevertheless, this way of representing patterns requires a subsequent analysis for the extraction of association rules. To solve this issue, it is possible to represent each specific item by means of two binary values in such a way that each of these pairs of values has a different meaning [39]. If the value is 00, it means that the specific item is included in the antecedent of the rule. The value 11 describes that the item is included in the consequent of the rule. Finally, the values 01 and 10 describe that the item is not considered to be included in the rule. Thus, this way of representing individuals requires a length of $2 \times n$, considering n as the number of distinct items in the database.

A different version of representing association rules by using a binary encoding was proposed in [40]. Here, each gene is represented as a binary value to denote whether a specific item is included in the rule or not. The main difference between this encoding criterion and the previously described is the use of an integer value to indicate the point in which the antecedent and consequent are separated. This integer value is placed into the first gene, so given a dataset comprising n different items, the chromosome length will be $n + 1$. Let us consider a sample dataset comprising ten different items, and the previous individual represented by the string of binary values 1001100101. Then, if the first gene is fixed to the value 3, it means that the three first items ($\{i_1, i_2, i_3\}$) will form the antecedent of the rule, whereas the rest of the items will describe the consequent, i.e. $\{i_4, i_5, \ldots, i_{10}\}$. Hence, the rule represented by this chromosome 31001100101 will be $\{i_1\} \rightarrow \{i_4, i_5, i_8, i_{10}\}$(see Fig. 4.1).

The growing interest in data storage has brought about the storage of much more information and, therefore, a higher number of items to be analysed. The use of datasets comprising an enormous number of items implies that extremely long strings of binary values are required to be used according to the previous proposals. These lengthy individuals hamper the mining process and the performance of the GAs may drop, so a variable-length encoding [40] is required. Besides, this way of representing individuals still suffers from continuous domains, requiring a previous discretization step to transform the continuous search space into a discrete one.

Considering the mining of continuous patterns as a hard process for any exhaustive search algorithm [2], first EAs in the pattern mining field were designed to solve this issue. A way of representing continuous patterns by means of GAs [25]

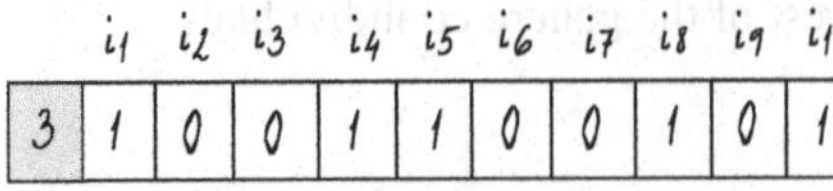

Fig. 4.1 Example of an individual in GA that represents the rule $\{i_1\} \rightarrow \{i_4, i_5, i_8, i_{10}\}$

Fig. 4.2 Example of an individual in GA that represents a set of continuous items or attributes

Fig. 4.3 Example of individual encoding to represent a varied set of continuous items

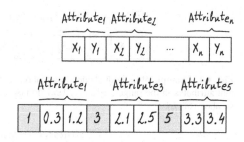

is to denote each individual as an encoding where each gene comprises two values to indicate the lower and upper bounds of a continuous interval. Thus, each individual is represented as string of length equal to $2 \times n$, considering n as the number of distinct items or attributes in the database, and each gene represents a specific attribute within the database. Figure 4.2 illustrates a sample individual encoding where both X_1 and Y_1 are the lower and upper bounds for the first attribute, respectively. Thus, the i-th gene in the string of genes (also known as chromosome) is equivalent to the i-th attribute that appears in the database. This way of representing patterns implies that the same set of items is always considered, so the aim is just to optimize the bounds of the set of numerical attributes. In situations where the aim is the looking for patterns with different sizes, the dataset needs to be modified just to include the desired set of items, and individuals will contain only the given items.

To solve the problem of mining continuous patterns that comprise different sets of items by just running a unique dataset, one of the easiest ways is to use a value to indicate whether the item is included or not. Thus, each item will require three different values: a binary value to indicate the presence or absence of the item, and two values to represents the upper and lower bounds of the interval. Nevertheless, this encoding criterion will require individuals of length $3 \times n$, considering n as the number of distinct items or attributes in the database. To reduce this length, some authors have proposed the use of an individual encoding that uses a variable-length string of values [26]. Similarly to the previous individual representation, each gene comprises three values. The main difference is that the first value is an integer value that indicate the item represented by the gene, so the chromosome illustrated in Fig. 4.3 indicates that the first attribute is defined in the range [0.3, 1.2], the third attribute is defined in the range [2.1, 2.5], and the fifth one in the range [3.3, 3.4]. Note that, according to this encoding criterion, the maximum number of items (attributes) to be included in any individual will be determined by the number of different items available in the dataset.

GAs have been used not only for mining continuous patterns but also association rules defined in a continuous domain. In this regard, some authors [4] have considered the individual representation as strings of genes where the i-th gene denotes the i-th attribute in the database, and each gene includes three different values. The first value of each gene describes whether the specific gene will be included in the rule or not, and in which part of the rule will be placed (antecedent

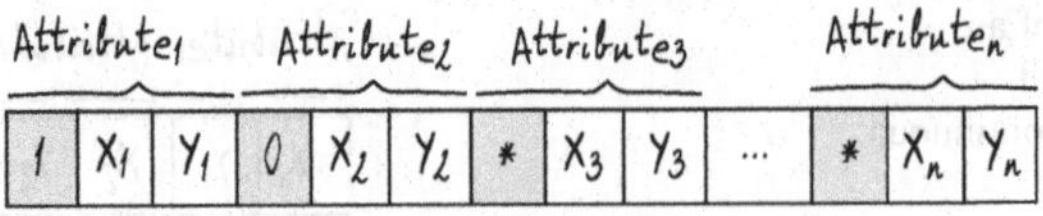

Fig. 4.4 Example of individual encoding that represents the rule $Att_2[X_2, Y_2] \rightarrow Att_1[X_1, Y_1]$. Note that attributes in the range $\{Att_3, \ldots, Att_n\}$ do not appear in the rule since they are marked by the symbol *asterisk*

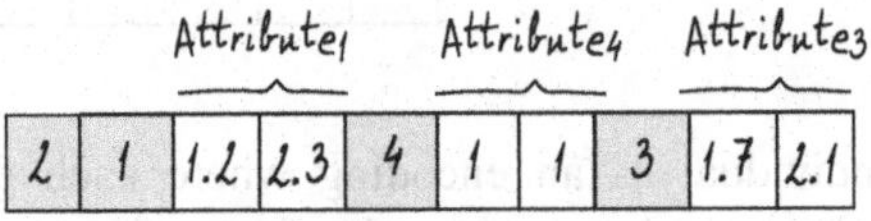

Fig. 4.5 Example of an individual in GA that represents a rule comprising three attributes: $Att_1[1.2, 2.3] \wedge Att_4 = 1 \rightarrow Att_3[0.7, 1.1]$

of consequent). If the value for the first part of the gene is 0, then it means that this item will be included in the antecedent of the rule. If the value is 1, then this item will be included in the consequent of the rule. Finally, if the value is *, then it means that this item will not be included in the rule. Thus, all the genes that have the value 0 on their first parts will form the antecedent of the rule while all the genes that have the value 1 will form the consequent of the rule. As for the second and third parts of each gene, they represent the lower and upper bounds of each interval. Figure 4.4 illustrates a sample individual by following this encoding criterion. Any attribute in the range $\{Att_3, \ldots, Att_n\}$ is marked by the symbol *, so none of these attributes will be included in the rule.

Considering the aforementioned encoding criterion and a dataset comprising n different attributes, then each individual is represented as string of values of length $3 \times n$. For the sake of reducing the length of the chromosomes, Yan et al. [40] proposed a novel encoding criterion to represent association rules that comprises either discrete and continuous patterns [32]. Each individual is encoded by means of a variable-length string of genes, so different number of items can be considered by each rule. Additionally, in order to determine which set of items will be form the antecedent of the rule, and which set of items will be included in the consequent, this encoding criterion uses the first gene as marker. Thus, a value of two will indicate that the two first attributes represented in the chromosome will be included in the antecedent, whereas the remain will form the consequent.

Analysing the individual representation described by Yan et al. [40], each gene is represented by means of three different values: the index of the attribute considered, and the lower and upper bounds of the continuous interval. In such a situation where the specific item within the rule is defined in a discrete domain, then the same value is assigned to the two genes reserved to represent the lower and upper bounds. For a better understanding, Fig. 4.5 illustrates a sample individual that represents a rule having three different attributes, i.e. $Att_1[1.2, 2.3] \wedge Att_4 = 1 \rightarrow Att_3[0.7, 1.1]$. Since the lower and upper interval values for Att_4 are the same, it means that this attribute is defined into a discrete domain, and it takes the value 1.

A major drawback of this way of representing individuals is that two individuals with different chromosomes can represent the same rule, so no rule can be uniquely represented. Many permutations of attributes in either the antecedent part or the consequent part of a chromosome can produce the same rule. For example, the rule described in Fig. 4.5, i.e. $Att_1[1.2, 2.3] \wedge Att_4 = 1 \rightarrow Att_3[0.7, 1.1]$, can be described by the following two chromosomes:

- $(2, (Att_1, 1.2, 2.3), (Att_4, 1, 1), (Att_3, 0.7, 1.1))$
- $(2, (Att_4, 1, 1), (Att_1, 1.2, 2.3), (Att_3, 0.7, 1.1))$

Association rules is more and more a useful method to identify hidden behaviour in large datasets, denoting how subsets of items influence the existence of others [3]. Most algorithms in pattern mining and, specially, in association rule mining are focused on the mining of positive patterns. Nevertheless, negative associations are also a major issue in the analysis of data, and can be used in situations where data comprise a large number of infrequent patterns. Negative association rules [20] describe negative relationship between items, i.e. the presence of a set of items is highly related to the absence of others. The mining of this type of associations is not trivial, and the search space is even much more bigger than the one of positive relations, so any algorithm might take more time to obtain results. Hence, the use of optimization techniques like GAs [30] can bring about interesting features, specially in computational and memory requirements.

The mining of negative and continuous association rules by means of GAs has been proposed by some authors [4], which described the individual representation as fixed-length strings of genes where the i-th gene denotes the i-th attribute in the database. Each of these genes are formed by four different values, so for a dataset comprising n different items defined into a continuous domain, any individual will be formed by a string of $4 \times n$ values. The first value of each gene is a binary value that is related to the description of whether the specific item is negative (a value 0 is used) or positive (a value 1). The second value of each gene determines whether the specific gene will be included in the rule or not, and in which part of the rule will be placed (antecedent of consequent). This second value for each gene is an integer value in the range [0, 2], so a value of 0 determines that the item will not be included in the rule. A value of 1 describes that the specific item will be included in the antecedent of the rule. Finally, if the value is 2, then it means that the item will be included in the consequent of the rule (see Fig. 4.6).

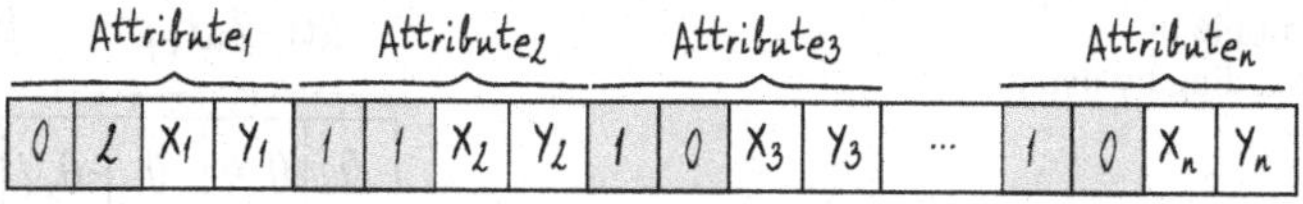

Fig. 4.6 Example of an individual in GA that represents the rule $Att_2[X_2, Y_2] \rightarrow \neg Att_1[X_1, Y_1]$. Note that attributes in the range $\{Att_3, \ldots, Att_n\}$ do not appear in the rule since they have the value 0 in the second value of the gene

Some authors described that the mining of continuous patterns have different problems caused by the sharp boundary of the intervals, which might not be intuitive for humans [10]. This problem can be overcomed by introducing fuzziness to the patterns [8], so the use of fuzzy sets to describe association rules facilitates the interpretation of rules in linguistic terms. The synergy between fuzzy logic and evolutionary algorithms provides useful tools for the analysis of patterns and the extraction of fuzzy association rules [9].

The use of fuzzy association rules is based on the fuzzy partition of the continuous domain [19]. This can be considered as a discretization approach where we establish a membership degree to the items [16], considering an overlapping between them. As a matter of example, let us consider a continuous attribute that represent the weight of a person. Considering three different labels, it is possible to associate a fuzzy partition to this domain as illustrated in Fig. 4.7, where the membership functions are labeled as L, M and H mean low, medium and high, respectively. The symbolic translation of a linguistic term is a number within the interval $[-0.5, 0.5)$ that expresses the domain of a linguistic label when it is moving between its two lateral labels. Thus, for an item represented as $Weight(M, 0.4)$, its meaning is that the weight is higher than Medium. On the contrary, the item represented as $Weight(H, -0.1)$ represents the weight a bit smaller than High.

Considering the mining of fuzzy association rules as a GA, Alcala et al. [5] proposed a model in which individuals are encoded by using a fixed-length string of real values. The size of this chromosome (or string of values) is $n \times m$, n is considered as the number of different items in the dataset, and m is defined as the number of linguistic labels per item. Each of the values comprises the displacements of the different linguistic labels for each item so given two different items with the aforementioned linguistic labels (L, M, and H), Fig. 4.8 shows a sample individual representation where the goal is to optimize the membership functions.

Fig. 4.7 Example of fuzzy partition for a continuous item using the labels L (low), M (medium) and H (high)

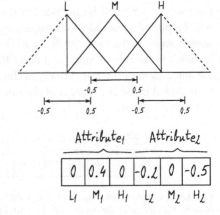

Fig. 4.8 Example of individual encoding by using fuzzy sets

4.2.2 Genetic Operators

As described in Chap. 3, a major feature of any EA is that the better the fitness value of an individual, the more often this individual is selected to breed and to pass its characteristics to later generations [11]. To emulate the breed process, some authors have considered those crossover and mutation operators traditionally used in GAs [17], whereas others have proposed specific operators according to their encoding criterion.

In the process of generating new individuals describing patterns of interest, Mata et al. [25] proposed a model where individuals were represented as fixed-length chains of genes. In this model, each gene comprises two values that denote the lower and upper bound of a specific attribute or item. Considering this individual representation, the simplest crossover operator that can be applied to the candidate solutions is the single point crossover [14]. This genetic operator takes two individuals from the set of parents and cut their string of values at some randomly chosen position. The new generated individuals, which have the same length, comprise the first genes (until the cut point) of the first individual and the following genes of the second individual (starting from the cut point), and vice versa. As for the mutation operator, *Mata et al.* described a mutation operator that acts on a single gene of a specific individual [25]. This mutation operator enables either to increase or decrease the intervals represented by each gene by shifting the lower (or upper) bound of such interval.

Another interesting crossover and mutation operators for continuous patterns were proposed as a way of mixing the bounds of the two parents [32]. The crossover operator consists in taking two individuals, called parents, at random and generating new individuals: for each attribute the interval is either inherited from one of the parents or formed by mixing the bounds of the two parents. Let us considered the domain of a continuous attribute, given two sample intervals for the continuous attribute, it is possible to generate four different intervals as shown Fig. 4.9. The first two intervals are the same as the parents. The last two new intervals are defined by using the lower bound of one parent and the upper bound of the other parent.

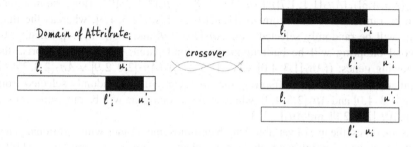

Fig. 4.9 Example of crossover operator defined by the QuantMiner algorithm [32]

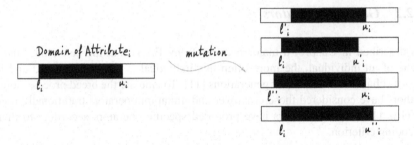

Fig. 4.10 Example of mutation operator defined by the QuantMiner algorithm [32]

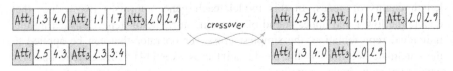

Fig. 4.11 Example of genetic operator, where two variable-length chromosomes are crossed to obtain two new individuals

As for the mutation operators proposed in [32], it works on a single individual and increases or decreases the lower or upper bound of its intervals. The main feature of this genetic operator is that the new bound is fixed to discard (if the interval width is decreased) or to involve (if the interval width is increased) no more than 10 % of transactions already covered by the original interval (see Fig. 4.10).

The aforementioned genetic operators were designed to work with individuals represented by a fixed-length chromosome. Nevertheless, this type of chromosomes might not be appropriate in the pattern mining field since patterns of different lengths can be found in any database. In such situations where the individuals of a population do not have the same length, the use of traditional crossover operators is not an option. Mata et al. [26] proposed a crossover operator to generate offspring whose genes were randomly chosen from one of the parents. Let us consider the two sample individuals depicted in Fig. 4.11, the first offspring will be built upon the first parent and will comprise the following genes: $((Att_1[1.3, 4.0] \vee Att_1[2.5, 4.3]), (Att_2[1.1, 1.7]), (Att_3[2.0, 2.9] \vee Att_3[2.3, 3.4]))$. Thus, the first gene will be randomly chosen from $Att_1[1.3, 4.0]$ and $Att_1[2.5, 4.3]$, whereas the third gene will be randomly selected from $Att_3[2.0, 2.9]$ and $Att_3[2.3, 3.4]$. Finally, the second offspring will be built upon the second parent, so it will comprise the following genes: $((Att_1[1.3, 4.0] \vee Att_1[2.5, 4.3]), (Att_3[2.0, 2.9] \vee Att_3[2.3, 3.4]))$. Similarly to the previous offspring, the first gene will be randomly selected from $(Att_1[1.3, 4.0]$ and $Att_1[2.5, 4.3]$, whereas the second gene will be randomly chosen from $Att_3[2.0, 2.9]$ and $Att_3[2.3, 3.4]$.

Sometimes, the use of variable-length chromosomes comes with additional genes that determine the point in which the antecedent and consequent are separated [40]. In general, the use of genes that act as markers of the antecedent/consequent parts is considered by using the first gene within the chromosome as an integer value.

Fig. 4.12 Example of genetic operator, where two variable-length chromosomes including an invariable gene (the first one) are crossed to obtain two new individuals

When a crossover genetic operator is applied to this type of individuals to obtain offspring, it is required to maintain the original structure, so the first gene should remain the same. Considering the aforementioned crossover operator, it is possible to cross two individuals by not dealing with the first gene (see Fig. 4.12), which represents the markers of the antecedent/consequent parts. Similarly, the use of a mutation operator implies that any gene (comprising three different values: the index of the attribute considered, and the lower and upper bounds of the continuous interval) except for the first gene (marker of the antecedent/consequent part) can be randomly changed by a different one. In such a way, the individual encoding is consistent and feasible offspring are obtained.

Alcala et al. [5] proposed a crossover operator based on the concept of neighbourhood, which allow the offspring genes to be around the genes of one parent or around a wide zone determined by the genes of both parents. This genetic operator presents two interesting features [21]. First, the operator assigns more probability for creating offspring near parents than anywhere in the search space. Second, the degree of diversity induced may be easily adjusted by means of the variation of the α parameter. Thus, the greater the α values, the higher the diversity introduced into the population.

The aforementioned crossover operator works as follows. Let us assume that $P' = \{p'_1, p'_2, \ldots, p'_n\}$ and $P'' = \{p''_1, p''_2, \ldots, p''_n\}$ are two parents defined into a real-coded domain, and each p'_i (or p''_i) is defined as an interval $[l_i, u_i]$ where l_i and u_i are the lower and upper bounds of the i-th gene. From these two parents, we obtain two offspring defined as follows:

- $O_1 = \{o_{11}, o_{12}, \ldots, o_{1n}\}$, where o_{1i} is a randomly chosen number from the interval $[l^1_i, u^1_i]$, with $l^1_i = max\{l_i, p'_i - I_i \cdot \alpha\}$, $u^1_i = min\{u_i, p'_i + I_i \cdot \alpha\}$, and $I_i = |p'_i - p''_i|$.
- $O_2 = \{o_{21}, o_{22}, \ldots, o_{2n}\}$, where o_{2i} is a randomly chosen number from the interval $[l^2_i, u^2_i]$, with $l^2_i = max\{l_i, p''_i - I_i \cdot \alpha\}$, $u^2_i = min\{u_i, p''_i + I_i \cdot \alpha\}$, and $I_i = |p'_i - p''_i|$.

4.2.3 Fitness Function

The fitness function is considered as one of the major elements in any EA, stating how close a given individual or candidate solution is to achieve the proposed aim. This function is highly dependent on the problem to be solved and it needs to be properly defined to accurately determine how close each individual is to the optimal solution. Focusing on the pattern mining problem, many authors have considered the use of interestingness measures to quantify how well the set of discovered patterns represents any important property of data. These measures can be divided into two different groups [38]: objective and subjective metrics. According to objective interestingness measures, patterns represent major features of data so their interest should be quantified by considering metrics that determine how representative a specific pattern is for the dataset. On the contrary, subjective interestingness measures requires any expert in the domain to quantify how promising the discovered knowledge is.

In general, most GAs proposed in the literature include fitness functions that are mainly based on objective quality metrics. As described in Chap. 1, all the objective quality measures are defined by using the number of transactions satisfied by each pattern or by a subset of the pattern. Given a pattern P defined as a subset of all the items $I = \{i_1, i_2, \ldots, i_n\}$ in a dataset, i.e. $P \subseteq I$, then it is possible to assert that the pattern P satisfies the transaction t_j defined from the set of all data transactions $T = \{t_1, t_2, \ldots, t_m\}$ if and only if $\{P \subseteq t_j : t_j \subseteq I\}$. Additionally, authors [2] determine that the number of transactions satisfied by the pattern P is defined as $f(P) = |\{\forall t_j \in T : P \subseteq t_j\}|$.

Using the number of transactions satisfied by a pattern, the GENAR (GENetic Association Rules) algorithm proposed a really interesting fitness function. The aim of this algorithm is to maximize the set of transactions satisfied by an individual (a pattern) and which has not been satisfied by any other individual yet. In each generation of the evolutionary process, GENAR checks the set of individuals and marks each dataset transaction as satisfied if at least one individual in the population satisfied the transaction. Authors of the GENAR algorithm determined a fitness function that depends on both the number of transactions $f(P)$ satisfied by the pattern P and the number of transactions in P that were not previously satisfied by any other pattern P', i.e. $f_{cov}(P) = |\{\forall t_j \in T : P \subseteq t_j\} \setminus \{\forall t_j \in T : P' \subseteq t_j\}|$. Thus, the goal is to maximize the following fitness function: $fitness(P) = f(P) - f_{cov}(P)$. As a matter of example, let us consider that the set of transactions satisfied by P is given by the set $\{t_1, t_3, t_4, t_6\}$, i.e. $f(P) = 4$; whereas the set of transactions $\{t_3, t_4, t_6\}$ determines transactions that were previously satisfied by other pattern P'. According to this information, it is possible to calculate $f_{cov}(P) = 1$. Hence, the fitness value for the individual that represents the pattern P is calculated as $fitness(P) = 4 - 1 = 3$.

Considering the same fitness function, it is also possible to include a penalization factor (PF) that quantifies the importance of covering transactions that were already covered by some previous patterns. The new fitness function, which is defined as $fitness(P) = f(P) - (f_{cov}(P) \times PF)$, includes the PF parameter that takes its values from the interval $(0, 1)$. PF values close to 0 imply that the general number of transactions satisfied is much more important, whereas values close to 1 determine that it is really important to satisfied transactions that were not previously satisfied by any other pattern.

A different fitness function was proposed in QuantMiner [32] to quantify the interest of the association rules represented by each individual in the GA. This fitness function is defined as the gain measure, which is based on both the confidence and the support of the consequent (see Chap. 2). Since the fitness function is based on the confidence measure, it is possible to consider a minimum user-specified confidence value so the aim is to optimize the gain measure by just considering individuals whose confidence value is higher than the threshold. Another GA that uses objective interestingness measures as fitness functions is EARMGA (Expanding Association Rule Mining Genetic Algorithm) [40], which includes certainty factor (CF) as the quality measure to determine the interest of the rule represented by each individual. Finally, a different fitness function [29] to quantify individuals that represent association rules in GAs is based on the support of the rule and the support of the antecedent. The aim of this fitness function is to find association rules that both their support and confidence are larger than the other rules.

Some authors have considered the use of subjective interestingness measures, which require any expert in the domain to quantify how promising the discovered knowledge is. This type of measures, together with objective quality measures, give rise to really interesting fitness functions. For instance, the amplitude of the intervals in continuous patterns can be considered as a subjective metric that enables to optimize the range of values defined for each interval. As a matter of example, let us consider a fitness function based on the number of transactions $f(P)$ satisfied by the pattern P. Using this value as a fitness function gives rise to extremely huge intervals that satisfy a large number of transactions. To solve this issue, it is required to include some subjective quality measures like the amplitude of the intervals. A really interesting function in this sense was included in the GAR (Genetic Association Rules) algorithm [26]. In this proposal, the authors included the amplitude as a parameter to penalize extremely huge intervals. Thus, if two patterns (comprising the same set of items) satisfy the same set of transactions, then the one whose intervals are smaller will provide the best information. The amplitude value for a set of items is calculated as the average amplitude for the set of items included in each specific pattern.

In some other situations, the size of the patterns discovered might be of interest, and many authors have considered that the lower the number of items in a pattern, the more comprehensible the pattern is. In other situations, the looking for extremely large patterns is a dare [42]. The size of patterns is highly related to the comprehensibility, which was described as a subjective measure in Chap. 2 in the sense that any pattern can be little comprehensible for a specific user and, at the same time, very comprehensible for a different one. Nevertheless, most of the authors [2] have determined that this metric can be defined as an objective quality measure where the fewer the number of items in a pattern, the more comprehensible the pattern is.

The two aforementioned measures (amplitude and number of items) are considered by the GAR algorithm [26], which describes a fitness value for a pattern P as follows: $fitness(P) = f(P) - (f_{cov}(P) \times PF) - (amplitude \times AF) + (\#items \times NF)$.

Similarly to the penalization factor *(PF)*, *AF* and *NF* are two factors that quantifies the importance of the amplitude and number of items in a pattern, respectively. Thus, this fitness function is able to optimize both objective and subjective interestingness measures at time.

Based on the fitness function used by the GAR algorithm, Alatas et al. [4] proposed a similar fitness function that incorporates both support and confidence as good quality measures to determine whether the rule is of interest or not. Additionally, the proposed fitness function includes two subjective quality measures such as the comprehensibility, i.e. the number of items, and the average amplitude of the intervals. The fitness value for each individual that represents a rule R is defined as $fitness(R) = (support(R) \times \alpha_1) + (confidence(R) \times \alpha_2) - (amplitude \times \alpha_3) - (\#items \times \alpha_4)$. Four parameters ($\alpha_1$, α_2, α_3 and α_4) are considered to increase or decrease the effects of parts of the fitness function, and any of them takes values in the range $(0, 1)$.

4.3   Algorithmic Approaches

A large volume of transaction data is generated everyday in a different number of areas, which brings about the need for analysing and getting useful information from that data. Some authors determined that databases in different domains are dynamic, and they increase or decrease every day. To mine association between items in this type of databases, the authors [35] proposed the DMARG (Dynamic Mining of Association Rules using Genetic Algorithms) algorithm, which follows a traditional GA (see Fig. 4.13) over the incremental database. This algorithm represents the

---

Require: *population_size*   {Number of individuals to be considered in the population}
Ensure: *population*
 1: *population ← ∅*
 2: *parents ← ∅*
 3: *offspring ← ∅*
 4: *population_size ← n*
 5: *population ←* generate(*population_size*)
 6: **for** ∀*member ∈ population* **do**
 7:     evaluate(*member*)
 8: **end for**
 9: **while** termination condition is not reached **do**
10:     *parents ←* select(*population*)
11:     *offspring ←* apply genetic operators on the set *parents*
12:     **for** ∀*member ∈ offspring* **do**
13:         evaluate(*member*)
14:     **end for**
15:     *population ←* update(*population,offspring*)
16: **end while**

Fig. 4.13 Pseudo-code of a traditional GA

whole set of items as a string of binary values so the length of each individual is equal to the number of different items in the database. In each generation of the evolutionary process, a subset of individuals are selected by means of a roulette-wheel selection, and these individuals are crossed and mutated by means of two well-known genetic operators (single point crossover and a simple mutation that modifies certain genes). Once the algorithm finishes, the database is modified by an incremental operator and the GA is applied on the new database again.

The growing interest in data storage has brought about the storage of much more information and, therefore, a higher number of items to be analysed. Besides, more and more data are defined in continuous domains, and traditional pattern mining algorithms require a previous discretization step to transform the continuous search space into a discrete one. One of the first GAs for mining frequent continuous patterns is known as GENAR (GENetic Association Rules) [25]. This algorithm was proposed as a way of dealing with continuous patterns without requiring a preprocessing step for dividing continuous domains into a fixed number of discrete intervals. In order to represent candidate solutions, GENAR uses an encoding criterion where each gene represents a specific continuous item or attribute within the database, and each of these genes comprise two values that indicate the lower and upper bounds of the continuous interval. Since patterns in a database are represented with different sizes, GENAR uses an iterative rule learning model [15] that returns only the best set of individuals of size n per run (see Fig. 4.14). Thus, the evolutionary procedure is run as many times as patterns the user wants to obtain, considering different sizes in each run.

Require: *maxRules, maxGenerations, population_size*
Ensure: *nRules*   {set of best individuals obtained}
1: *nRules* ← 0
2: **while** *nRules* < *maxRules* **do**
3:     *nGenerations* ← 0
4:     *population* ← generate(*population_size*)
5:     **for** ∀*member* ∈ *population* **do**
6:         evaluate(*member*)
7:     **end for**
8:     **while** *nGenerations* < *maxGenerations* **do**
9:         *parents* ← select(*population*)
10:        *offspring* ← apply genetic operators on the set *parents*
11:        **for** ∀*member* ∈ *offspring* **do**
12:            evaluate(*member*)
13:        **end for**
14:        *population* ← update(*population, offspring*)
15:        *nGenerations* ← *nGenerations* + 1
16:    **end while**
17:    return the best rule in *population*
18:    *nRules* ← *nRules* + 1
19: **end while**

Fig. 4.14 Pseudo-code of the GENAR algorithm [25]

In each generation of GENAR, two genetic operators are used to obtain new individuals. First, a single point crossover operator is carried out, which takes two individuals from the set of parents and cuts their string of values at some randomly chosen position. The new generated individuals comprise the first genes (until the cut point) of the first individual and the following genes of the second individual (starting from the cut point), and vice versa. Second, a mutation operator is used in GENAR to act on a single gene of a specific individual, increasing or decreasing the intervals represented by shifting the lower (or upper) bounds. Any of the new generated individuals are evaluated according to a fitness function that maximizes the set of transactions satisfied by each individual and which has not been satisfied by any other individual yet. This kind of fitness function was previously described in Sect. 4.2.3. Finally, it should be noted that GENAR enables association rules to be obtained from the set of patterns discovered. Each specific pattern (represented as a string of genes, one per attribute) represents an association rule where the consequent is defined by the last attribute (last gene), whereas the antecedent is defined by the remain attributes (all the genes except for the last one).

Another interesting approach for mining continuous patterns is known as GAR (Genetic Association Rules) [26], which is mainly based on GENAR [25]. The GAR algorithm represents each individual by means of a variable-length string of values that represents a specific pattern. The fact of representing individuals as variable-length strings of values implies that this algorithm is able to discover a varied set of patterns at once. Furthermore, GAR does not represent a single association rule per individual but a set of rules. Thus, for each individual of length k, which represents a pattern of interest, it is possible to obtain $2^k - 2$ different association rules.

Similarly to the GAR algorithm [26], first algorithms for mining association rules were based on the discovery of frequent patterns and, in a subsequent process, the extraction of association rules from the previously extracted patterns [41]. To solve this issue, the QuantMiner [32] algorithm (see Fig. 4.15) is a GA for mining association rules in a direct way on continuous domains, not requiring to generate frequent patterns in a previous step. It works by optimizing the intervals of a set of rules previously mined, which are considered as a set of rule templates. A rule template is a predefined association rule, either chosen by the user or computed by the system, and it is used as starting point for the mining process. The GA looks for the best intervals for the numeric attributes included in the rule template.

Another algorithm to extract rules in a direct way as proposed by Alatas et al. [4]. In this algorithm (see Fig. 4.16), the initial population is generated by using a really interesting methodology, which avoids the problems related to those algorithms that create the initial population randomly. The use of methods to generate the initial population randomly might produce some undesiderable situations such as the generation of infeasible individuals, the creation of a set of individuals in a zone of the search space that produces local solutions and far to the global solution, or even a set of similar individuals that are placed in the nearest neighborhood.

In order to avoid some of the undesiderable situations described above, Alatas et al. [4] proposed a method that enables dissimilar individuals to be generated for the initial population. Let us consider a chromosome that represents a pattern as

Require: *maxGenerations*   {maximum number of iterations to be considered}
Ensure: *solutions*   {set of best individuals obtained}
  1: $R \leftarrow$ set of rule templates chosen by the user or computed by the system
  2: *solutions* $\leftarrow \emptyset$
  3: **for** $\forall r \in R$ **do**
  4:     *population* $\leftarrow$ generate a random population of size n using the template r
  5:     *parents* $\leftarrow \emptyset$
  6:     *offspring* $\leftarrow \emptyset$
  7:     *bestIndividual* $\leftarrow \emptyset$
  8:     *currentGen* $\leftarrow 0$
  9:     **for** $\forall member \in population$ **do**
 10:         evaluate(*member*)
 11:     **end for**
 12:     **while** *currentGen* $\leq$ *maxGenerations* **do**
 13:         *parents* $\leftarrow$ select(*population*)
 14:         *offspring* $\leftarrow$ apply genetic operators on the set *parents*
 15:         **for** $\forall member \in offspring$ **do**
 16:             evaluate(*member*)
 17:         **end for**
 18:         *population* $\leftarrow$ update(*population*,*offspring*)
 19:         *bestIndividual* $\leftarrow$ keep the best individual in *population*
 20:     **end while**
 21:     *solutions* $\cup$ *bestIndividual*
 22: **end for**
 23: return *solutions*

Fig. 4.15 Pseudo-code of the QuantMiner algorithm [32]

a vector of genes (one per item in the database) that comprises a value 1 or 0 to denote if the specific item appears or not in the pattern, respectively. If a specific pattern $\{i_1, i_3, i_4, i_6, i_7\}$ is represented for a dataset comprising 10 different items, another pattern $\{i_2, i_5, i_8, i_9, i_{10}\}$ that is inversion of the previous one can also be created. Thus, two dissimilar chromosomes (patterns) can be represented for each individual randomly created, as described by Alatas et al. [4]. As an extension of the aforementioned way of creating dissimilar individuals, it is also possible to divide the randomly created chromosome, which acts as a baseline, into two equal parts to obtain three new chromosomes. First, the inversion of the first part is taken to obtain a new chromosome. Second, the new chromosome is created by taking the inversion of the second part. Finally, the third chromosome is produced by using the inversion of the whole chromosome that acts as baseline. Thus, it is possible to create four dissimilar individuals when creating just one. Considering the same aforementioned example for a dataset comprising 10 different items, two equal parts can be used to generate individuals from a specific one. The two parts will be formed from item i_1 to item i_5, and from item i_6 to item i_{10}. Thus, taking the pattern $\{i_1, i_3, i_4, i_6, i_7\}$ as a baseline, it is possible to create the following patterns: $\{i_2, i_6, i_7\}$, $\{i_1, i_3, i_4, i_8, i_9, i_{10}\}$ and $\{i_2, i_5, i_8, i_9, i_{10}\}$. Note that for n-dividing points, then the number of derived chromosomes from one randomly generated is equal to $2^n - 1$.

Require: *maxGenerations, population_size*   {maximum number of iterations and individuals}
Ensure: *population*   {set of best individuals obtained}
 1: *nGenerations* ← 0
 2: *population* ← generate(*population_size*)
 3: **while** *nGenerations* < *maxGenerations* **do**
 4:    **while** size of *population* < 2 ∗ *population_size* **do**
 5:       select a genetic operator
 6:       **if** selected operator is mutator **then**
 7:          *offspring* ← apply an adaptive mutation on the set *population*
 8:       **else**
 9:          *parents* ← select(*population*)
10:          *offspring* ← apply crossover operator on the set *parents*
11:       **end if**
12:       *population* ← *population* ∪ *offspring*
13:    **end while**
14:    **for** ∀*member* ∈ *population* **do**
15:       evaluate(*member*)
16:    **end for**
17:    *population* ← remove redundant solutions from the set *population*
18:    *population* ← update(*population, population_size*)
19:    *nGenerations* ← *nGenerations* + 1
20: **end while**
21: return the set of rules in *population*

Fig. 4.16 Pseudo-code of the algorithm proposed by Alatas et al. [4]

Following with the algorithm proposed by Alatas et al. [4], it is interesting to note that it was proposed not only as a way of dealing with continuous patterns but also with both positive and negative patterns. This algorithm represents the individuals as a string of genes where each of these genes comprises four different parts. The first value of each gene is a binary value that is related to the description of whether the specific item is negative (a value 0 is used) or positive (a value 1). The second value of each gene determines whether the specific gene will be included in the rule or not, and in which part of the rule will be placed (antecedent of consequent). Finally, the third and fourth parts of each gene are used to represent the lower and upper bounds of each interval.

As in any GA, all the generated individuals are evaluated according to a fitness function. Since the algorithm proposed by Alatas et al. [4] enables association rules to be mined directly, the fitness function is based on four quality measures (as described in Sect. 4.2.3) used to quantify the interest of any association rule. First, two objective quality measures such as support and confidence are considered in the fitness function. Second, the fitness function also includes two subjective quality measures such as the comprehensibility, i.e. the number of items, and the average amplitude of the intervals.

One of the first GAs that does not consider support/confidence as metrics to be optimized was proposed by Yan et al. [40]. Instead, this algorithm, known as EARMGA (Expanding Association Rule Mining Genetic Algorithm), includes certainty factor (CF) as the quality measure to determine the interest of the rule represented by each individual. CF [37] calculates the variation of the probability of the consequent of a rule when consider only those transactions satisfied by the antecedent.

Additionally, one of the main features of the EARMGA algorithm is its encoding criterion, which enables individuals to be represented by means of variable length chromosomes. This way of representing individuals allows association rules comprising different numbers of items to be obtained. According to the encoding criterion, the first gene is reserved to indicate the point in which the antecedent and consequent are separated. Then, a series of genes is included by considering each gene as a set of three different values: the index of the attribute considered, and the lower and upper bounds of the continuous interval.

Finally, an algorithm for mining fuzzy association rules was proposed by Alcala et al. [5]. This algorithm (see Fig. 4.17) enables the optimization and mining of both the membership functions and fuzzy association rules in continuous domains. The algorithm receives the database, the number of predefined linguistic labels, the number of maximum evaluations and the population size. The algorithm is run for a predefined number of evaluations, and each iteration optimizes both the membership functions and the fuzzy association rules defined on the membership functions. Thus, individuals are encoded by using a fixed-length string of real values where the length of each individual is $n \times m$, n is considered as the number of different items in the dataset, and m is defined as the number of linguistic labels per item. Each of the real values that form the chromosome denotes the displacements of the different linguistic labels for each item.

Require: *maxEvaluations, population_size* {maximum number of evaluations and individuals}
Ensure: *solutions* {set of best individuals obtained}
1: *parents* ← ∅
2: *offspring* ← ∅
3: *solutions* ← ∅
4: *valuation* ← ∅
5: *population* ← generate(*population_size*)
6: **for** ∀*member* ∈ *population* **do**
7:     evaluate(*member*)
8: **end for**
9: **while** *evaluation* ≤ *maxEvaluations* **do**
10:     *parents* ← select(*population*)
11:     *offspring* ← apply genetic operators on the set *parents*
12:     **for** ∀*member* ∈ *offspring* **do**
13:         evaluate(*member*)
14:     **end for**
15:     *evaluation* ← update the number of evaluations carried out
16:     *population* ← update(*population,offspring*)
17:     *solutions* ← update the set of best solutions found
18: **end while**
19: return *solutions*

Fig. 4.17 Pseudo-code of the algorithm proposed by Alcala et al. [5]

4.4 Successful Applications

Market basket analysis was the first application domain in which pattern mining and association rules were applied. Since its definition in the early 1990s [3], the pattern mining task has provoked tremendous curiosity among experts on different application fields [31, 33]. In many of these fields, authors have considered the use of GAs to solve the problem under study. For instance, medical domains have been analysed by means of a GA that extracts interesting relationships between attributes [18].

In the medical analysis mentioned above, the authors considered two medical databases comprising information about breast cancer and cervix uteri cancer. They used the GAR algorithm [26], which represents each individual by means of a variable-length string of values that represents a specific pattern. The resulting set of rules discovered comprises association rules that are satisfied by a huge number of transactions. For example, 92 % of the transactions satisfied that patients *"having protein nm23"* no more than six, *"survived"* at least 3 years. In the analysis, many rules can be considered as trivial, but others are of high interest. For example, *"period of survive"* is larger than 5 years when the there was no *"renewal"* and *"relatives"* free of the cancer.

A different field where the mining of relationships between items is of high interest is in modeling ozone series from satellite observations. The extraction of relationships between meteorological variables is an important topic related to climate change. Some authors [24] have considered the analysis of total ozone content series modeling in the Iberian Peninsula using association rules obtained by an evolutionary algorithm. Experimental results have been carried out on data measured at three locations (Lisbon, Madrid and Murcia) of the Iberian Peninsula. As prediction variables for the association rules the authors have considered several meteorological variables, such as outgoing long-wave radiation, temperature at 50 hPa level, tropopause height, and wind vertical velocity component at 200 hPa. The results obtained in the analysis were compared to the results obtained by different methodologies, and the results are quite similar, denoting the interest of the proposed approach for the future in similar problems.

Climatological time series have been typically analysed by means of statistical methods. Nevertheless, as some authors have described [22], the good performance and simplicity presented by these methods in synthetic data cannot be achieved when applying to real-world time series. In this regard, the authors proposed a GA to be applied to discover association rules between temperature, wind and ozone time series from June 2003 to September 2003. Results were compared to the well-known Apriori algorithm, obtaining an error value lower when the new approach is applied.

Marketing-oriented firms are especially concerned with modeling consumer behavior to improve their information and aid their decision processes on markets, so marketing experts apply complex models and statistical methodologies to infer conclusions from data. Orriols-Puig et al. [28] proposed an approach based on the discovery of fuzzy relationships between patterns by means of GAs. The main advantage with regard to traditional methodologies used by marketing experts is that no a priori knowledge is required by the proposed approach.

Finally, another field in which the mining of association rules by means of GAs has been applied to is in microarray analysis [23], describing interrelations between genes. The associations inferred by the proposed approach were also inferred by different methods proposed in the literature when the same dataset was used. In total, thirteen rules previously inferred by other proposals were also obtained, with addition of new seven rules discovered only by the proposed GA. Authors stated that the biological relevance of the rules inferred by the approach was verified by analyzing whether such rules reflect functional properties relating to the different cell-cycle phase.

References

1. M. Affenzeller, S. Winkler, S. Wagner, and A. Beham. *Genetic Algorithms and Genetic Programming: Modern Concepts and Practical Applications.* Chapman & Hall/CRC, 1st edition, 2009.
2. C. C. Aggarwal and J. Han. *Frequent Pattern Mining.* Springer International Publishing, 2014.
3. R. Agrawal, T. Imielinski, and A. N. Swami. Mining association rules between sets of items in large databases. In *Proceedings of the 1993 ACM SIGMOD International Conference on Management of Data,* SIGMOD Conference '93, pages 207–216, Washington, DC, USA, 1993.
4. B. Alatas and E. Akin. An efficient genetic algorithm for automated mining of both positive and negative quantitative association rules. *Soft Computing,* 10(3):230–237, 2006.
5. J. Alcala-Fdez, R. Alcala, M. J. Gacto, and F. Herrera. Learning the membership function contexts for mining fuzzy association rules by using genetic algorithms. *Fuzzy Sets and Systems,* 160(7):905–921, 2009.
6. Michael J. Berry and Gordon Linoff. *Data Mining Techniques: For Marketing, Sales, and Customer Support.* John Wiley & Sons, Inc., New York, NY, USA, 2011.
7. M. Bessaou and P. Siarry. A genetic algorithm with real-value coding to optimize multimodal continuous functions. *Structural and Multidisciplinary Optimization,* 23(1):63–74, 2002.
8. M. J. del Jesús, J. A. Gámez, P. González, and J. M. Puerta. On the discovery of association rules by means of evolutionary algorithms. *Wiley Interdisciplinary Reviews: Data Mining and Knowledge Discovery,* 1(5):397–415, 2011.
9. M. Delgado, N. Marín, D. Sánchez, and M. A. Vila. Fuzzy association rules: general model and applications. *IEEE Transactions on Fuzzy Systems,* 11:214–225, 2003.
10. D. Dubois, H. Prade, and T. Sudkamp. On the representation, measurement, and discovery of fuzzy associations. *IEEE Transactions on Fuzzy Systems,* 13(2):250–262, 2005.
11. A. A. Freitas. *Data Mining and Knowledge Discovery with Evolutionary Algorithms.* Springer-Verlag Berlin Heidelberg, 2002.
12. M. Gendreau and J. Potvin. *Handbook of Metaheuristics.* Springer Publishing Company, Incorporated, 2nd edition, 2010.
13. B. Goethals. Survey on Frequent Pattern Mining. Technical report, Technical report, HIIT Basic Research Unit, Department of Computer Science, University of Helsinki, Finland, 2003.
14. D. E. Goldberg. *Genetic Algorithms in Search, Optimization and Machine Learning.* Addison-Wesley Longman Publishing Co., Inc., Boston, MA, USA, 1st edition, 1989.

15. A. González and F. Herrera. Multi-stage genetic fuzzy systems based on the iterative rule learning approach. *Mathware & soft computing*, 4(3):233–249, 1997.
16. F. Herrera. Genetic fuzzy systems: taxonomy, current research trends and prospects. *Evolutionary Intelligence*, 1(1):27–46, 2008.
17. J. H. Holland. *Adaptation in Natural and Artificial Systems*. The University of Michigan Press, 1975.
18. H. Kwasnicka and K. Switalski. Discovery of association rules from medical data: classical and evolutionary approaches. *Annales UMCS, Informatica*, 4(1):204–217, 2006.
19. Y. Lee, T. Hong, and W. Lin. Mining fuzzy association rules with multiple minimum supports using maximum constraints. In M. Negoita, R. Howlett, and L. Jain, editors, *Knowledge-Based Intelligent Information and Engineering Systems*, volume 3214 of *Lecture Notes in Computer Science*, pages 1283–1290. Springer Berlin Heidelberg, 2004.
20. Y. Li, A. Algarni, and N. Zhong. Mining Positive and Negative Patterns for Relevance Feature Discovery. In *Proceedings of the 16th ACM SIGKDD International Conference on Knowledge Discovery and Data Mining*, KDD '10, pages 753–762, Washington, DC, USA, 2010. ACM.
21. M. Lozano, F. Herrera, N. Krasnogor, and D. Molina. Real-coded memetic algorithms with crossover hill-climbing. *Evolutionary Computation*, pages 3–273, 2004.
22. M. Martinez-Ballesteros, F. Martinez-Alvarez, A. Troncoso, and J. C. Riquelme. Quantitative association rules applied to climatological time series forecasting. In *Proceedings of the 10th International Conference on Intelligent Data Engineering and Automated Learning, IDEAL 2009*, pages 284–291, Brugos, Spain, 2009.
23. M. Martinez-Ballesteros, I. A. Nepomuceno-Chamorro, and J. C. Riquelme. Inferring gene-gene associations from quantitative association rules. In *Proceedings of the 11th International Conference on Intelligent Systems Design and Applications*, ISDA 2011, pages 1241–1246, Cordoba, Spain, 2011.
24. M. Martinez-Ballesteros, S. Salcedo-Sanz, J. C. Riquelme, C. Casanova-Mateo, and J. L. Camacho. Evolutionary association rules for total ozone content modeling from satellite observations. *Chemometrics and Intelligent Laboratory Systems*, 109(2):217–227, 2011.
25. J. Mata, J. L. Alvarez, and J. C. Riquelme. Mining numeric association rules with genetic algorithms. In *Proceedings of the 5th International Conference on Artificial Neural Networks and Genetic Algorithms*, ICANNGA 2001, pages 264–267, Taipei, Taiwan, 2001.
26. J. Mata, J. L. Alvarez, and J. C. Riquelme. Discovering numeric association rules via evolutionary algorithm. In *Proceedings of the 6th Pacific-Asia Conference on Advances in Knowledge Discovery and Data Mining*, PAKDD 2002, pages 40–51, Taipei, Taiwan, 2002.
27. M. Mitchell. *An Introduction to Genetic Algorithms*. MIT Press, Cambridge, MA, USA, 1998.
28. A. Orriols-Puig, J. Casillas, and F. J. Martínez-López. Unsupervised Learning of Fuzzy Association Rules for Consumer Behavior Modeling. *Mathware & soft computing*, 16(-):29–43, 2009.
29. H. R. Qodmanan, M. Nasiri, and B. Minaei-Bidgoli. Multi objective association rule mining with genetic algorithm without specifying minimum support and minimum confidence. *Expert Systems with Applications*, 38:288–298, 2011.
30. N. S. Rai, S. Jain, and A. Jain. Mining Interesting Positive and Negative Association Rule Based on Improved Genetic Algorithm (MIPNAR_GA). *International Journal of Advanced Computer Science and Applications*, 5(1), 2014.
31. C. Romero and S. Ventura. Educational data mining: a review of the state of the art. *IEEE Transactions on Systems, Man, and Cybernetics, Part C*, 40(6):601–618, 2010.
32. A. Salleb-Aouissi, C. Vrain, and C. Nortet. QuantMiner: A Genetic Algorithm for Mining Quantitative Association Rules. In *Proceedings of the 20th International Joint Conference on Artificial Intelligence*, IJCAI'97, pages 1035–1040, Hyderabad, India, 2007.
33. D. Sánchez, J. M. Serrano, L. Cerda, and M. A. Vila. Association Rules Applied to Credit Card Fraud Detection. *Expert systems with applications*, (36):3630–3640, 2008.
34. A. Savasere, E. Omiecinski, and S. B. Navathe. An efficient algorithm for mining association rules in large databases. In *Proceedings of the 21th International Conference on Very Large Data Bases*, VLDB '95, pages 432–444, San Francisco, CA, USA, 1995.

35. P. D. Shenoy, K. G. Srinivasa, K. R. Venugopal, and L. M. Patnaik. Evolutionary approach for mining association rules on dynamic databases. In *Proceedings of the 7th Pacific-Asia Conference on Advances in Knowledge Discovery and Data Mining*, PAKDD 2003, pages 325–336, Seoul, Korea, 2003.

36. P. D. Shenoy, K. G. Srinivasa, K. R. Venugopal, and L. M. Patnaik. Dynamic association rule mining using genetic algorithms. *Intelligent Data Analysis*, 9(5):439–453, 2005.

37. E. H. Shortliffe and B. G. Buchanan. A model of inexact reasoning in medicine. *Mathematical biosciences*, 23:351–379, 1975.

38. P. Tan and V. Kumar. Interestingness Measures for Association Patterns: A Perspective. In *Proceedings of the Workshop on Postprocessing in Machine Learning and Data Mining*, KDD '00, New York, USA, 2000.

39. P. P. Wakabi-Waiswa and V. Baryamureeba. Extraction of interesting association rules using genetic algorithms. *International Journal of Computing and ICT Research*, 2(1):1818–1828, 2008.

40. X. Yan, C. Zhang, and S. Zhang. Genetic algorithm-based strategy for identifying association rules without specifying actual minimum support. *Expert Systems with Applications*, 36:3066 – 3076, 2009.

41. C. Zhang and S. Zhang. *Association rule mining: models and algorithms*. Springer Berlin / Heidelberg, 2002.

42. F. Zhu, X. Yan, J. Han, P. S. Yu, and H. Cheng. Mining colossal frequent patterns by core pattern fusion. In *Proceedings of the IEEE 23rd International Conference on Data Engineering*, ICDE 2007, pages 706–71, Istanbul, Turkey, 2007. IEEE.

Chapter 5
Genetic Programming in Pattern Mining

Abstract This chapter describes the use of genetic programming for the mining of patterns of interest and the extraction of accurate relationships between patterns. The current chapter first describes the canonical representation of genetic programming and the use of grammars to restrict the search space. Then, it describes different approaches based on genetic programming for mining association rules of interest, paying special attention to the grammars used to restrict the search space, the genetic operators applied and the fitness functions considered by different approaches. Finally, this chapter deals with a series of application domains in which the use of genetic programming for mining association rules has been a successfully applied.

5.1 Introduction

With the rapid growth in size and number of available databases, the extraction of knowledge, regularities or high level information from data has become essential to comprehend some behaviour and to support decision taking. Pattern mining [7] and the discovery of relationships between patterns are considered as useful tasks for the extraction of comprehensible, useful, and non-trivial knowledge from large and, sometimes, messy databases. According to some authors [1], the pattern mining task can be described as a really time consuming process that requires large amount of main memory. In pattern mining, the search space exponentially grows with the number of single items or singletons in data. Given a dataset comprising the set of items $I = \{i_1, i_2, \ldots, i_n\}$, then $2^{|I|} - 1$ different patterns can be obtained, so the complexity of the mining task is highly dependent on the number of singletons.

As it is described by many authors [4, 12], and detailed in the previous chapter, genetic algorithms (GAs) have emerged as robust and practical optimization methods [6] that can be applied to different data domains. The use of GAs is considered by some authors as a way of optimizing and looking for patterns of interest [27, 39], so thanks to this type of algorithms, the pattern mining problem can be considered as a combinatorial optimization problem where solutions are represented as strings of genes and each gene describes the way in which a specific item is analysed. Thus, the use of GAs for mining patterns of interest has come out as a really interesting methodology [3], obtaining accurate results and performing well in terms of runtime and scalability.

© Springer International Publishing Switzerland 2016

S. Ventura, J.M. Luna, *Pattern Mining with Evolutionary Algorithms*,

DOI 10.1007/978-3-319-33858-3_5

Despite the fact that GAs have achieved a really good performance in the pattern mining field, the way in which their solutions are encoded hampers the mining process since it does not allow to restrict the search space. Most GAs in this field look for patterns of interest in the whole search space, and some minor restrictions can be considered in terms of size of the patterns. Existing proposals based on GAs do not consider external knowledge to be used, so the subjectivity is only considered by means of some subjective quality measures like, for instance, comprehensibility [4].

Considering the restriction of the search space as a baseline, the use of grammars to include syntax constraints is a really interesting way to reduce the search cost. Nevertheless, the use of grammars comes with the consequent risk of not reaching the optimal solutions due to the grammars' constraints [28]. Grammars in pattern mining have been successfully applied to different approaches [22, 24], reducing the search space since it can be really extensive in this field. Additionally, grammars can be considered as a useful way of introducing subjective knowledge to the pattern mining process, enabling patterns of interest to be obtained by users with different goals. Grammars are highly related to the background knowledge of the user [9], and they allow to define solutions that have flexible and expressive structures that ease the interpretability of the extracted knowledge.

The use of grammars has been profitably applied to genetic programming (GP), which is defined as an evolutionary and very flexible heuristic technique that represents the solutions to the problem by means of trees, enabling any kind of function and domain knowledge to be considered [10]. Solutions represented as trees include two kind of symbols: leaves and internal nodes. Leave nodes correspond to variables and constants, whereas internal nodes correspond to operators and functions. Grammar-guided genetic programming (GGGP or G3P) [38] enables different data types to be manipulated. G3P employs a context-free grammar to generate any feasible solution to the problem under study [28], and the grammar constrains the search space and solutions are generated by applying a set of productions rules.

Since grammars can be modelled according to the problem under study, different structures of the patterns, types of items and logical operators have been considered by numerous approaches. The first use of grammars for mining patterns of interest and, specially, association rules, was considered by the G3PARM algorithm [21]. This algorithm defines a grammar where solutions comprise a set of items in the antecedent of the rule, whereas the consequent is formed only by a single item. This grammar enables both positive and negative patterns to be obtained, and only a simple modification in the grammar provokes the extraction of either positive or negative patterns. Thanks to G3PARM many different approaches based on grammars have been proposed in recent years [22, 24].

Analysing the features of G3P, a really important one is its ability to introduce subjective knowledge into the pattern mining process. G3P enables the structure of a pattern or the form of an association rule to be defined by the end user according to his/her preferences. This freedom to organise and define the type of knowledge that the end user wants to obtain is highly important in many application domains, and the educational environment [34] is one of them. Algorithms for mining patterns of interest have been successfully applied to a wide range of educational problems and tasks [33] such as making recommendations for students and teachers, student modelling, predicting student performance, etc. A really interesting educational application in which patterns of interest where discovered by means of G3P was recently proposed [35]. The aim of this work was to help instructors in decision making about improving a course, so the conclusions reached after improving the course were really hopeful since students enhanced their results [35].

5.2 General Issues

As previously described, the use of GP and, specially, G3P [28] enables any feasible solution to the problem under study to be generated. A major feature of G3P is that it includes a grammar to generate GP individuals, and this grammar constraints either the search space and the solutions. In this section, both GP and G3P are formally described, denoting different individual representations and genetic operators.

5.2.1 Canonical Genetic Programming

Genetic programming (GP) [2] was defined as an evolutionary computation methodology that automatically solves different problems without any knowledge about how to do it. Individuals in GP are defined as computer programs that evolve through generations, so during the process, GP constructs new programs by applying genetic operations on the computer programs. Some authors [4] have considered that GP is a paradigm of evolutionary algorithms where both the individual representation and the corresponding genetic operators are its main features.

According to the individual representation, individuals that represent computer programs are usually expressed as tree structures rather than strings of values. First, leave nodes correspond to terminal symbols, which comprise variables and constants. Second, internal nodes describe operators and functions. In general, individuals in GP usually grow in size and shape unlike conventional GAs that use a fixed-length string of values. Nevertheless, there are some GAs that use a variable-length representation and there are also some GP algorithms that use some limit of the tree size [4]. Thus, it is possible to assert that the main difference between both GA and GP is that the latter can contain not only values but also functions. Figure 5.1 shows a sample tree representation of function expressed as $min(4*x+y, x*x*2)$.

It is common in the GP literature to represent expressions in a Polish notation also known as prefix notation, so the expression for adding the numbers 3 and 8 is written as (+ 3 8) rather than (3 + 8). This notation often makes it easier to analyse the relationship between (sub)expressions and their corresponding (sub)trees. Considering the sample individual illustrated in Fig. 5.1, the prefix

Fig. 5.1 Example of an individual representation by means of a tree shape

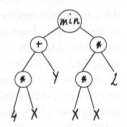

notation is given by the expression $(min\ (+\ (*\ 4\ x)\ y)\ (*\ (*\ x\ x)\ 2))$. It should be noted that the performance of the GP trees will obviously depend on the way in which GP trees are implemented. In some environments, tree-based representations may be too memory-inefficient since they require the storage and management of numerous pointers (from node to node of the tree structure). Besides, in cases where individuals are represented as prefix notations, brackets might become redundant and trees may be simpler.

Focusing on the generation of individuals, it is a stochastic procedure like in other EAs found in the literature. Many authors have considered different approaches to obtain random individuals in GP [31], and two of the first methodologies in this regard are known as full and grow methods. In both methods a maximum predefined depth is used, so no extremely large trees can be generated. The depth of a specific node within the tree is defined as the number of nodes needed to be traversed to reach it (starting from the root node, whose depth is 0). Similarly, the depth of a tree is defined as the largest depth that can be found in it by considering its leaves or terminal symbols. Considering the sample individual shown in Fig. 5.1, the depth of the tree is 3, whereas the depth of the node that represents the sum function ($+$) is 1.

The full method generates trees having leaves which are at the same depth. In this method, nodes are taken at random from the set of functions until the maximum tree depth is reached and, at this point, only terminal nodes can be chosen. Notice that the fact that all the leaves have the same depth does not necessarily imply that all the generated trees have the same shape. In fact, it only happens if all the functions have an equal arity, also known as the number of subtrees. Figure 5.2 illustrates the process in which a full tree of depth 2 is created. The full tree represents the function $min(4 + y, x * 7)$, and all the functions have an arity of 2.

As for the grow method, it enables a great variety of sizes and shapes to be generated. Unlike full method, the grow method can randomly choose any node from either the set of functions or the set of terminal values. The only requirement to be considered in this method is the maximum depth of the tree, so only terminal symbols can be selected once the maximum depth is reached. A sample individual that was obtained by the aforementioned grow method is illustrated in Fig. 5.3, which considers the value of 2 as the maximum depth of the tree. The sample resulting tree obtained by the grow method represents the function $min(5 + x,\ 7)$. An additional example of a tree obtained by the grow method is the one showed in Fig. 5.1, which maximum depth is equal to 3. Finally, noted that the shape of the individuals obtained by this method is highly dependent on the type of the

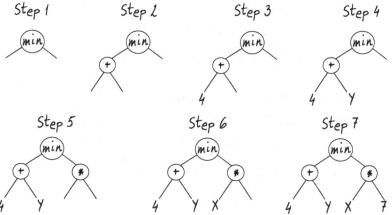

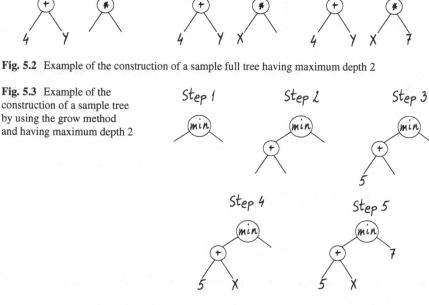

Fig. 5.2 Example of the construction of a sample full tree having maximum depth 2

Fig. 5.3 Example of the construction of a sample tree by using the grow method and having maximum depth 2

functions and, more specifically, on the number of subtrees that each function generates. Additionally, the size of the trees is highly related to the number of terminal symbols, so high number implies the generation of relatively short trees, whereas a low number implies that the grow method tends to similarly behave to the full one.

Since GP is considered by many authors as an extension of GAs [31], GP also uses the two main genetic operators that enable new solutions to be produced. The first of these genetic operators is crossover, which usually takes randomly chosen subtrees from the parents to obtain a (or more than one) new individual. A typical crossover operator in GP lies in taking two parents and two randomly chosen subtrees of the parents to obtain a new individual. This individual is generated by replacing a subtree of one of the parents with a subtree of the second parent, as illustrated in Fig. 5.4. This way of creating new individuals can be selected

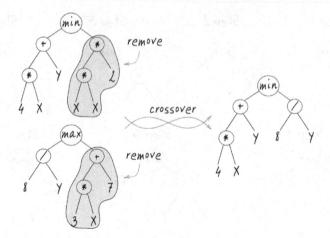

Fig. 5.4 Example of a traditional GP crossover where a new individual is obtained from two parents

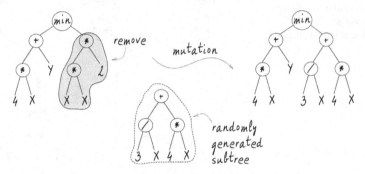

Fig. 5.5 Example of a simple GP mutation where a subtree is replaced by another randomly generated

multiple times, so two parents might take part in the creation of multiple offspring by choosing and discarding different subtrees.

In GP, most of the proposed mutation genetic operators are based on the replacement of subtrees of a parent by a new subtree randomly generated (see Fig. 5.5). Another type of mutation operator in GP is to replace a randomly chosen subtree by another subtree with the following features: (1) the new subtree has the same depth; (2) the depth of the new subtree has a smaller depth; or (3) the new depth is bigger than the previous one. Other form of mutation replaces a single node in the tree with another node of the same type. Thus, a function will be replaced by another randomly-chosen function, whereas a leave node will be replaced by another randomly-generated leave node. Notice that if the node selected to be mutated is a leave, then both the depth and shape of the offspring are equal to those of the parent. On the contrary, if the selected node is an internal node, then both the depth and shape of the resulting tree may vary. Finally, noted that there are different mutation operators described as new forms of mutation in GP, which obtained promising results in finding solutions [5, 20].

5.2.2   Syntax-Restricted Programming

Grammars can be considered as one of the core representation structures in computer science, and they have become popular as methods for representing restrictions on general domains, limiting the expressions obtained in GP [18]. Grammars were successfully applied to GP, giving rise to grammar-guided genetic programming (GGGP or G3P) [38], which employs a grammar to generate any feasible solution to the problem under study, and that grammar enables the search space to be constrained. According to McKay et al. [28], there is no G3P system that was firstly proposed but a set of them that were independently implemented at about the same time. One of these systems was proposed by Whigham et al. [37] in which a context-free grammar (CFG) was used to generate individuals defined by means derivation trees obtained from the grammar.

The use of a CFG in GP allowed the introduction of the concept of language in this field, so the G3P process might result in a reduction in the search space. Individuals in G3P are derivation trees that represent solutions belonging to the language defined by the CFG, so G3P always generates valid individuals that belong to the search space. An additional major feature of any G3P proposal is its ability to introduce subjective knowledge, enabling solutions to be obtained by users with different goals. Noted that grammars are highly related to the background knowledge of the user, and solutions obtained by means of grammars are flexible and expressive, which eases the interpretability of the extracted knowledge. Similarly to GP, in G3P each individual is represented by means of both a genotype and a phenotype. The former is defined as a derivation syntax tree based on the defined language, whereas the latter refers to the meaning of the tree structure (the expression).

A grammar G is defined as a four-tuple $(\Sigma_N, \Sigma_T, P, S)$ where Σ_T represents the alphabet of terminal symbols and Σ_N the alphabet of non-terminal symbols. It can be noted that they have no common elements, i.e., $\Sigma_N \cap \Sigma_T = \emptyset$. In order to encode an individual using a G3P approach, a number of production rules from the set P are applied starting from the start symbol S. A production rule is defined as $\alpha \rightarrow \beta$ where $\alpha \in \Sigma_N$, and $\beta \in \{\Sigma_T \cup \Sigma_N\}^*$. It should be noted that in any CFG there may appear the production rule $\alpha \rightarrow \varepsilon$, i.e., the empty symbol ε is directly derived from α. To obtain individuals, a number of production rules is applied from the set P, obtaining a derivation syntax tree for each individual, where internal nodes contain only non-terminal symbols, and leaves contain only terminals.

For a better understanding, let us consider a sample grammar G as shown in Fig. 5.6. To obtain a derivation syntax tree for each individual, a number of production rules is applied from the set P, where internal nodes contain only

$$G = (\Sigma_N, \Sigma_T, P, S) \text{ with:}$$

$$
\begin{aligned}
S &= \text{Min} \\
\Sigma_N &= \{\text{Min, Operator, Symbol, Value}\} \\
\Sigma_T &= \{\text{'+', '−', '*', '/', '1', '2', '3', '4', '5', '6', '7', '8', '9'}\} \\
P &= \{\text{Min = Operator, Operator ;} \\
&\quad \text{Operator = Symbol, Operator, Operator | Value ;} \\
&\quad \text{Symbol = '+' | '−' | '*' | '/' ;} \\
&\quad \text{Value = '1' | '2' | '3' | '4' | '5' | '6' | '7' | '8' | '9' ;}\}
\end{aligned}
$$

Fig. 5.6 Sample context-free grammar expressed in extended BNF notation

Fig. 5.7 Sample derivation syntax tree generated from the sample context-free grammar (see Fig. 5.6)

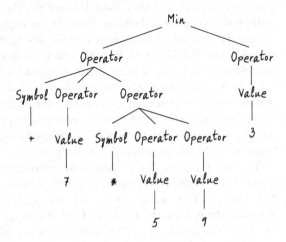

non-terminal symbols, and leaves contain only terminals. This process begins from the starting symbol *Min*, which always has a child node representing two expressions that can be a simple value or a complex expression. Thus, considering the aforementioned grammar *G*, the sample individual shown in Fig. 5.7 can be obtained. Analysing the derivation syntax tree as a prefix notation, the following expression is obtained (*Min (Operator Symbol + Operator 7 (Operator Symbol * Operator Value 5 Operator Value 9)) (Operator Value 3)*), which represents the function *min*(7 + (5 * 9), 3).

One of the main features of using grammars is the flexibility to represent solutions, so a simple change in the grammar is able to produce completely different solutions. As a matter of example, let us modify the previous CFG (see Fig. 5.6) to represent a function that obtains the maximum of two expressions. In such a case, it is enough to replace the starting symbol *Min* by *Max*. Additionally, let us consider that one of the expressions is the product of the variable *X* and a number in the range [1, 9]. Thus, to introduce these constraints, the aforementioned CFG is modified and a new one is obtained (see Fig. 5.8). A sample individual generated by using this new grammar is shown in Fig. 5.9, which represents the function *max*(7 + (5 * 9), *X* * 2). As graphically illustrated, a small change in the original grammar is able to produce completely new functions having special constraints, e.g. one of the expressions within the function is always formed by the variable *X* and a number in the range [1, 9].

$G = (\Sigma_N, \Sigma_T, P, S)$ with:

S = Max
Σ_N = {Max, Operator, Symbol, ProductSymbol, Variable, Value}
Σ_T = {'X', '+', '−', '*', '/', '1', '2', '3', '4', '5', '6', '7', '8', '9'}
P = {Max = Operator, Product ;
      Operator = Symbol, Operator, Operator | Value ;
      Product = ProductSymbol, Variable, Value ;
      ProductSymbol = '*' ;
      Symbol = '+' | '−' | '*' | '/' ;
      Variable = 'X' ;
      Value = '1' | '2' | '3' | '4' | '5' | '6' | '7' | '8' | '9' ;}

Fig. 5.8 Sample context-free grammar with some extra constraints

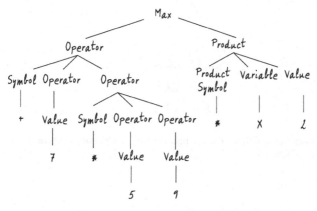

Fig. 5.9 Sample derivation syntax tree generated from the sample context-free grammar (see Fig. 5.8)

Focusing on the generation of new individuals by means of crossover and mutation genetic operators, G3P works similarly to canonical GP with the additional constraint that crossover points are required to have the same grammar label [28]. Figure 5.10 shows a sample crossover operator in G3P where subtrees with the same grammar labels are swapped. Finally, considering the mutation genetic operator in G3P, the new subtree will start by the same non-terminal symbol (see Fig. 5.11), so a completely new subtree is obtained by applying production rules to the aforementioned non-terminal symbol.

Finally, it is interesting to describe some general characteristics of G3P, which were analysed in depth by McKay et al. [28]. As previously described, one of the major characteristics of using grammars in GP is its ability to restrict the search space, which is one of the most important argument for the use of G3P. The use of grammars, usually provided in BNF notation or similar formalism, provides the ability to alter the search space simply by changing the input grammar. This ability to alter the search space is a really important issue in many application domains, where some subjective knowledge provided by the user should be included. Another

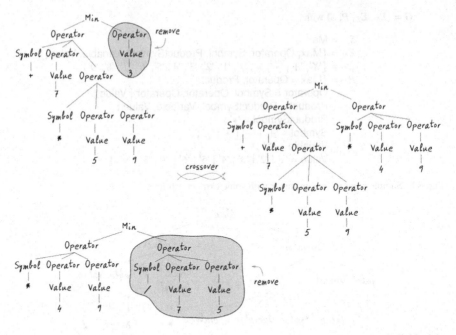

Fig. 5.10 Example of a G3P crossover restricted by the grammar where a new individual is obtained from two parents

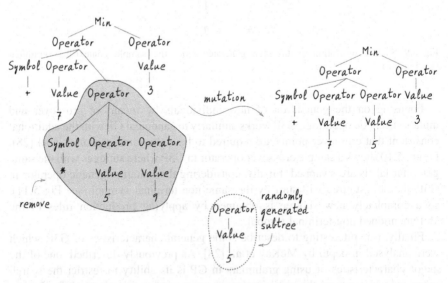

Fig. 5.11 Example of a G3P mutation restricted by the grammar where a subtree is replaced by another randomly generated

important characteristics of G3P is related to its genetic operators, which replace a component of a solution with another having similar functionality, reducing the likelihood of producing semantically meaningless solutions.

All the aforementioned characteristics denote the strength of using grammars in GP. However, the use of G3P is not extent of drawbacks. The design and use of genetic operators in both GP and G3P is not an easy task. Any new operator has to meet either the general requirements of the tree consistency and the additional constraints imposed by the grammar. In such situations where the genetic operators produce an infeasible individual, different mechanisms are required to repair it. The simplest way is just to discard the infeasible solution. Another possibility is to include a new method to repair the individual and to guarantee the feasibility of the solution. Finally, another G3P drawback that should be described is the recursivity of the grammar, which may produce extremely large individuals. To solve this issue, different methodologies that constraints either the depth of the trees or the number of derivations are considered.

5.3 Algorithmic Approaches

As described in previous section, grammars can be modelled according to the problem under study and the knowledge provided by the user. These characteristics are really interesting in the pattern mining field so different structures of the patterns, types of items and logical operators have been considered by numerous approaches. In this section, different approaches for mining both frequent and infrequent patterns are considered. Then, a really interesting approach for optimizing continuous patterns is described. Finally, the first G3P approach for mining association rules in relational databases is detailed.

5.3.1 Frequent Patterns

The first use of grammars for mining patterns of interest and, specially, association rules, was described by Luna et al. [21]. This algorithm, known as G3PARM (Grammar-Guided Genetic Programming for Association Rule Mining), defines a grammar where solutions comprise a set of items in the antecedent of the rule, whereas the consequent is formed only by a single item. This grammar enables both positive and negative patterns to be obtained, and only a simple modification in the grammar provokes the extraction of either positive or negative patterns.

G3PARM encodes each individual as a sentence of the language generated by the grammar G (see Fig. 5.12), considering a maximum number of derivations to avoid extremely large trees. Thus, considering the grammar defined in this approach, the following language is obtained $L(G) = \{$ *Comparison* $(\wedge$ *Comparison*$)^n \rightarrow$ *Comparison* $, n \geq 0\}$. To obtain individuals, a set of derivation steps is carried

$G = (\Sigma_N, \Sigma_T, P, S)$ with:

S = Rule
Σ_N = {Rule, Antecedent, Consequent, Comparison, Comparator, Attribute}
Σ_T = {'∧', '=', '≠', 'name', 'value'}
P = {Rule = Antecedent, Consequent ;
       Antecedent = Comparison | '∧', Comparison, Antecedent ;
       Consequent = Comparison ;
       Comparison = Comparator, Attribute ;
       Comparator = '=' | '≠' ;
       Attribute = 'name', 'value' ;}

Fig. 5.12 Context-free grammar defined for the G3PARM algorithm

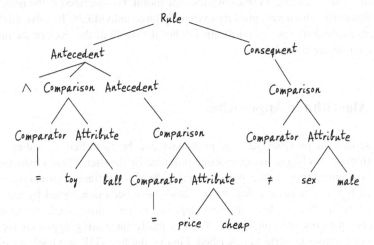

Fig. 5.13 Sample individual encoded by G3PARM by using the CFG shown in Fig. 5.12

out by applying the production rules declared in the set P and starting from the
start symbol *Rule*. It should be noted that each terminal symbol randomly chooses
the names (*name*) and values (*value*) from the dataset metadata, so considering
the sample metadata illustrated in Table 5.1, the terminal symbol *name* adopts
any value from the following: *toy*, *toy's price*, or *sex*. Once the symbol *name* is
assigned, a random value is selected in its domain, e.g., *ball*, *teddy*, or *doll* for
the attribute *toy*. Thus, each individual in G3PARM is represented by a syntax-
tree structure according to the defined grammar, and considering a maximum depth
to avoid infinite derivations. Figure 5.13 illustrates a sample individual encoded
from the grammar G (see Fig. 5.12) and the sample metadata (see Table 5.1).
The individual represented by the derivation syntax-tree represents the rule **IF**
toy = ball ∧ price = cheap **THEN** *sex ≠ male*.

Once the individuals are encoded by means of the CFG, the G3PARM algorithm
follows a traditional evolutionary schema (see Fig. 5.14) that uses an elite population
to keep the best individuals obtained along the evolutionary process. In order to

Table 5.1 Sample metadata

Attributes	Values
Toy	Ball, teddy, doll
Toy's price	Cheap, expensive
Sex	Male, female

Require: *population_size,maxGenerations* {Number of individuals and generations considered}
Ensure: *elite_population*
1: *population* ← ∅
2: *elite_population* ← ∅
3: *offspring* ← ∅
4: *number_generations* ← 0
5: *population* ← generate(*population_size*)
6: **for** ∀*member* ∈ *population* **do**
7: evaluate(*member*)
8: **end for**
9: **while** *number_generations* < *maxGenerations* **do**
10: *offspring* ← apply genetic operators on a set of parents selected from *population*
11: **for** ∀*member* ∈ *offspring* **do**
12:   evaluate(*member*)
13: **end for**
14: *population* ← update(*population, offspring, elite_population*)
15: *elite_population* ← best individuals from {*population, offspring, elite_population*}
16: *number_generations* + +
17: **end while**
18: return *elite_population*

Fig. 5.14 Pseudo-code of the G3PARM algorithm

obtain new individuals, G3PARM uses two genetic operators traditionally used in G3P. The crossover operator creates new individuals by exchanging two parent derivation subtrees from two randomly selected nodes (with the same label) in each of them. Additionally, the mutation operator randomly select a node within the tree and a completely new subtree is obtained by applying production rules to the node.

Finally, it should be noted that the aim of this algorithm is the extraction of highly frequent patterns and reliable association rules from that patterns. Thus, the algorithm uses a fitness function that is based on the support quality measure, so the aim of G3PARM is to maximize this function. Additionally, in order to obtain reliable associations, a minimum threshold value for the confidence quality measure is considered, so no rule having a confidence value lower than a predefined value can be included in the set *elite_population* as a solution. Once the algorithm finishes, the set *elite_population* comprises the set of best rules discovered along the evolutionary process.

As previously described, the grammar considered in the first G3P proposal for mining patterns of interest was based on the discovery of patterns on discrete domains, considering either positive and negative patterns, so continuous domains need to be preprocessed. Besides, this grammar was defined in such a way that association rules discovered just contain one item in the consequent. With the aim

$G = (\Sigma_N, \Sigma_T, P, S)$ with:

S = Rule
Σ_N = {Rule, Antecedent, Consequent, Comparison, Comparator_Discrete,
          Comparator_Continuous, Attribute_Discrete, Attribute_Continuous}
Σ_T = {'∧', '=', '≠', '<', '≤', '>', '≥', 'name', 'value'}
P = {Rule = Antecedent, Consequent ;
         Antecedent = Comparison | '∧', Comparison, Antecedent ;
         Consequent = Comparison | '∧', Comparison, Consequent ;
         Comparison = Discrete, Attribute | Continuous, Attribute ;
         Discrete = '≠' | '=' ;
         Continuous = '<' | '≤' | '>' | '≥' ;
         Attribute = 'name', 'value' ;}

Fig. 5.15 Context-free grammar defined for the G3PARM algorithm considering new features

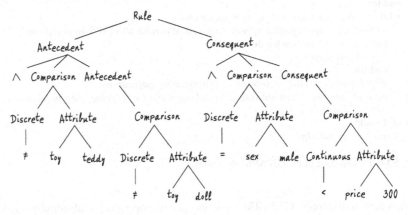

Fig. 5.16 Sample individual encoded by G3PARM by using the CFG shown in Fig. 5.15

of mining any type of association rule (the consequent might contain more than a single item) on different domains (either discrete and continuous), a new CFG was proposed in [22], which is illustrated in Fig. 5.15. Considering this new grammar G, the following language is obtained $L(G) = \{$ Comparison $(\wedge$ Comparison$)^n \rightarrow$ Comparison $(\wedge$ Comparison$)^m$, $n \geq 0, m \geq 0\}$. Thanks to language defined by this grammar, the G3PARM algorithm (see Fig. 5.14) is able to mine any type of association rule directly from data, not requiring any previous step to transform data or to create rules from a set of patterns of interest. Let us consider now that the attribute *toy's price* from the sample metadata shown in Table 5.1 is defined in a continuous domain in the range $[0, 1500]$. Figure 5.16 illustrates a sample individual encoded from the new grammar (see Fig. 5.15) and the sample metadata (see Table 6.1) with the continuous domain included. The individual represented by the derivation syntax-tree represents the rule **IF** *toy* $\neq$ *teddy* $\wedge$ *toy* $\neq$ *doll* **THEN**

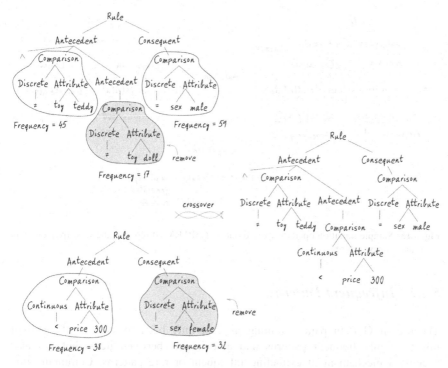

Fig. 5.17 Sample crossover operator carried out by G3PARM. It exchanges the least frequent item from a parent by the most frequent item from another parent

$sex = male \land price < 300$. As shown, this rule comprises either discrete and continuous items, and the consequent of the rule might comprise more than one item.

As for the genetic operators used in this new version of G3PARM [22], authors proposed a modified version of classical G3P operators described in the previous section. These new genetic operators do not randomly choose a subtree to be swapped or mutated, instead they are based on the fact that highly frequent items might produce frequent patterns and, therefore, frequent association rules. The main idea of these genetic operators is to replace the least frequent item defined within a parent to produce a new more frequent item. Thus, the crossover operator (see Fig. 5.17) removes the least frequent item from a parent and it is replaced by the most frequent item from the second parent. The same concept is considered for the mutation operator, which replaces the least frequent item from a parent to produce a completely new random subtree as illustrated in Fig. 5.18.

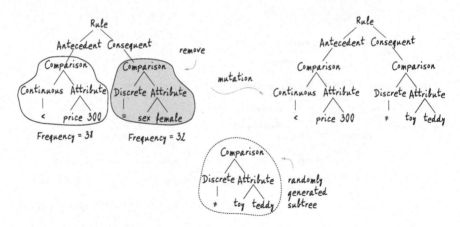

Fig. 5.18 Sample mutation operator carried out by G3PARM. It replaces the least frequent item to produce a new subtree

5.3.2   Infrequent Patterns

The use of G3P in pattern mining has not only been focused on the aim of mining highly frequent patterns and associations between items but has also become a mechanism of extracting infrequent or rare patterns. Communication failure detection [32], analysis of interesting rare patterns in telecommunication networks [19], recognition of patients who suffer a particular disease that does not often occur [30], or credit card fraud detection [36] are some of the applications where it is interesting to use association rules and, particularly, rare association rules.

The process of mining rare association rules generates might obtain two types of infrequent association rules as described in [16]. Given maximum support and minimum confidence thresholds, an association rule of the from $X \rightarrow Y$ is defined as perfectly rare association rule (PRAR) if the rule's confidence is greater than or equal to a minimum confidence threshold, the support of the entire rule is lower than a maximum predefined value, and the frequency of occurrence of each single item within the rule does not overcome the minimum support threshold (see Eq. (5.1)).

$$PRAR \begin{cases} Confidence(X \rightarrow Y) \geq mininimum\_confidence \\ Support(X \rightarrow Y) < maximum\_value \\ \forall i : i \in (X \cup Y), f(i) < maximum\_value \end{cases} \quad (5.1)$$

Additionally, considering the same threshold values for maximum support and minimum confidence, an association rule is defined as imperfectly rare association rule (IRAR) [17] if the rule's confidence is greater than or equal to a minimum confidence threshold, the support of the complete rule is lower than a maximum support threshold, and there is at least one item within the rule having a frequency of occurrence greater than or equal to the maximum support threshold (see Eq. (5.2)).

Require: *population_size, maxGenerations* {Number of individuals and generations considered}
Ensure: *elite_population*
1: *population* ← ∅
2: *elite_population* ← ∅
3: *offspring* ← ∅
4: *number_generations* ← 0
5: *population* ← generate(*population_size*)
6: **for** ∀*member* ∈ *population* **do**
7:     evaluate(*member*)
8: **end for**
9: **while** *number_generations* < *maxGenerations* **do**
10:     *offspring* ← apply genetic operators on a set of parents selected from *population*
11:     **for** ∀*member* ∈ *offspring* **do**
12:         evaluate(*member*)
13:     **end for**
14:     *population* ← update(*population, offspring, elite_population*)
15:     *aux_population* ← *population* ∪ *offspring* ∪ *elite_population*
16:     *elite_population* ← ∅
17:     **for** ∀*member* ∈ *population* **do**
18:         **if** *member* ∉ *elite_population* **then**
19:             **if** getFitness(*member*) > 0 **then**
20:                 **if** getConfidence(*member*) > *confidenceTreshold* **then**
21:                     **if** getLift(*member*) > 1 **then**
22:                         *elite_population* ← (*elite_population* ∪ *member*)
23:                     **end if**
24:                 **end if**
25:             **end if**
26:         **end if**
27:     **end for**
28:     *number_generations* + +
29: **end while**
30: return *elite_population*

Fig. 5.19 Pseudo-code of the Rare-G3PARM algorithm

$$IRAR \begin{cases} Confidence(X \to Y) \geq minimum\_confidence \\ Support(X \to Y) < maximum\_value \\ \exists i : i \in (X \cup Y), f(i) \geq maximum\_value \end{cases} \tag{5.2}$$

A really interesting algorithm for mining not only PRAR but also IRAR was proposed by Luna et al. [23]. In this proposal, known as Rare-G3PARM (see Fig. 5.19), the CFG used to encode the solution is the same as the one used by the modified version of G3PARM, so either continuous and discrete items can be included in the association rules discovered. Additionally, the use of this CFG (see Fig. 5.15) enables either positive and negative associations to be mined, and there is no restriction for the number of items in either the antecedent and consequent.

Rare-G3PARM looks for infrequent associations of high interest for the end user, and this interest is calculated by means of two objective quality measures as confidence and lift (see Chap. 2). The use of these quality measures introduces some changes in the process of choosing the best individuals for the elite population, so the algorithm does not follow the procedure of G3PARM but a completely new evolutionary process (see Fig. 5.19). A new individual is considered to be included in the elite population if and only if it satisfies some minimum quality thresholds for the confidence and lift metrics (see lines 23 and 24, Fig. 5.19). As for the fitness function, Rare-G3PARM defines the support quality measure as the fitness function to be minimized, but considering some restrictions, so four different fitness functions were proposed by the authors.

In order to understand the usefulness of the four proposed fitness functions, it should be noted that the use of pattern mining techniques in many application domains discover extremely infrequent patterns, and these patterns might cause noisy associations. In such situations, it is essential to correctly establish a boundary between infrequent and noisy associations. Besides, the distinction between infrequent and frequent association rules is also a hard process, and it depends on the subjectivity of the user. The aim of these four fitness functions considered by Rare-G3PARM is to allow the user to determine the boundaries and the type of function that he/she wants to use according to his/her previous knowledge.

The first fitness function (see Eq. (5.3) and Fig. 5.20) provides a maximum fitness value in the middle of a certain interval provided by a minimum and a maximum support value. The closer the support of a rule is to the interval limits, the lower its fitness value is. Out of this interval, a zero value is assigned. This first fitness function is useful for application domains with a large number of both noisy association rules and rules that are close to being considered as rare. In this way, a progressive fitness function is applied, since there is not a clear difference between noisy, rare, and frequent association rules. Therefore, the more distant the support value is from the predefined thresholds, the better is the rule.

Fig. 5.20 Fitness function that provides minimum and maximum support values. This fitness function is mathematically described in Eq. (5.3)

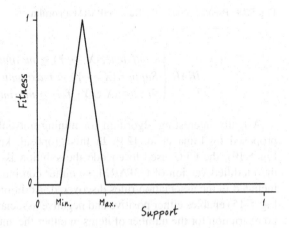

Fig. 5.21 Fitness function that provides minimum and maximum support values and solutions within this range of values are equally promising. This fitness function is mathematically described in Eq. (5.4)

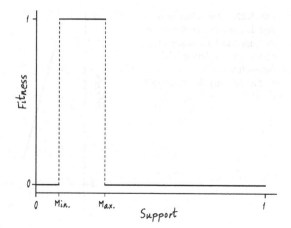

$$F_1(rule) = \begin{cases} \dfrac{Sup(rule) - Min}{\dfrac{(Max + Min)}{2}} & \text{if } Min \leq Sup(rule) \leq \dfrac{(Max + Min)}{2} \\[2em] \dfrac{Max - Sup(rule)}{\dfrac{(Max + Min)}{2}} & \text{if } \dfrac{(Max + Min)}{2} \leq Sup(rule) \leq Max \\[2em] 0 & \text{Otherwise} \end{cases} \quad (5.3)$$

The second fitness function, defined in Eq. (5.4) and Fig. 5.21, is based on the fact that any association rule with a support value within the predefined range is equally promising, so it is essential to establish a mechanism to properly differentiate among rules, bringing the confidence and lift measures into play. It should be noted that the main application domains of this fitness function are those where there is a clear difference between noisy and rare association rules.

$$F_2(rule) = \begin{cases} 1 & \text{if } Min \leq Sup(rule) \leq Max \\ 0 & \text{Otherwise} \end{cases} \quad (5.4)$$

Finally, the other two fitness functions (see Eqs. (5.5) and (5.6)) were defined to minimize and maximize the support within a given interval, respectively. One of these fitness functions (see Fig. 5.22) assigns a fitness value based on the proximity of the support to the lower limit. This fitness function is specific for domains where there is a clear difference between noisy and rare association rules but the difference between rare and frequent rules is not that clear. The other fitness function (see Fig. 5.23) maximizes the fitness value when the support of the rule is closer to the upper limit of the interval. This fitness function allows of differentiating between frequent and rare association rules in situations where there is a clear limit between both types.

Fig. 5.22 Fitness function
that determines that the lower
the support of a solution the
better is that solution. This
fitness function is
mathematically described in
Eq. (5.5)

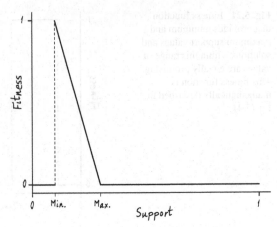

Fig. 5.23 Fitness function
that determines that the lower
the support of a solution the
better is that solution. This
fitness function is
mathematically described in
Eq. (5.6)

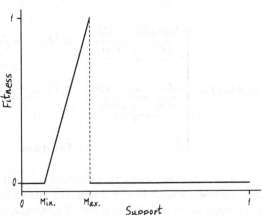

$$F_3(rule) = \begin{cases} 1 + \dfrac{Min - Sup(rule)}{Max - Min} & \text{if } Min \leq Sup(rule) \leq Max \\ 0 & \text{Otherwise} \end{cases} \quad (5.5)$$

$$F_4(rule) = \begin{cases} \dfrac{Sup(rule) - Min}{Max - Min} & \text{if } Min \leq Sup(rule) \leq Max \\ 0 & \text{Otherwise} \end{cases} \quad (5.6)$$

As for the genetic operators used by Rare-G3PARM to generate new individuals
along the evolutionary process, the same idea as G3PARM is considered. The main
difference lies in the item to be replaced or removed. As previously described,
G3PARM aims to replace the least frequent item defined within a parent to produce
a new more frequent item. Thus, the crossover operator removes the least frequent
item from a parent and it is replaced by the most frequent item from the second
parent. The same concept is considered for the mutation operator, which replaces

the least frequent item from a parent to produce a completely new random subtree. In this regard, Rare-G3PARM works similarly except for the fact that its aim is to replace the most frequent item defined within a parent to produce a new one that appears in data with a lower frequency. Thus, the crossover operator defined in Rare-G3PARM removes the most frequent item from a parent and it is replaced by the least frequent item from another parent, whereas the mutation operator replaces the most frequent item from a parent to produce a completely new random subtree.

5.3.3 *Highly Optimized Continuous Patterns*

The use of G3P has not only been used for mining frequent and infrequent association rules, but also for mining highly optimized association rules defined in continuous domains. Noted that the optimisation of the continuous values is also a challenge, searching for rules that do not comprise patterns that represent unnecessary ranges of values [11]. Besides, the task of searching for the optimal parameter values in any evolutionary algorithm is an arduous process and unsuccessful results might be obtained. To solve these two issues, a G3P model was proposed [24], which is defined as a free-parameter algorithm that self-adapts to the required parameters, meaning that it is highly useful for users with no experience in evolutionary computation, and which searches for highly representative continuous patterns.

The proposed approach to optimize continuous association rules includes a process where the support and confidence of each rule is not the only point of interest, but also the distribution of transactions satisfied by the rule. According to the authors in [24], the aim of mining highly optimised association rules in continuous domains is to select the right amount of values to contain as few gaps as possible. A gap is defined as a space which does not comprise any transaction. For a better understanding, let us consider a sample range $[A, B]$ of continuous items (see Fig. 5.24). From the search space of this item, the sample interval X can be obtained in the looking for frequent patterns, comprising 16 transactions. However, analysing the distribution of the transactions within the interval, it is discovered that there is a huge gap in the middle of such interval, so it can be split into two sub-intervals X_1 and X_2. These new intervals have a lower support value but they present a higher interest since their transactions are uniformly distributed. Thus, the aim of this algorithm is to look for the right width of values, that is, values containing as few blank spaces as possible.

Fig. 5.24 Sample continuous interval to demonstrate the importance of avoiding gaps in the looking for optimal range of values

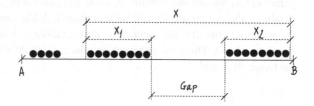

Fig. 5.25 Representation of the F_1 function, which denotes the importance of reducing the gaps

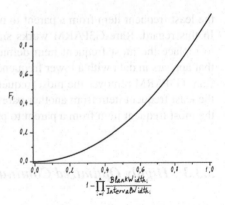

In order to calculate the quality of an association rule r, the proposed G3P approach uses the combination of three different functions $fitness(r) = F_1 + F_2 \times F_3$, which are described as follows. F_1 (see Eq. (5.7)) is responsible for searching for a set of transactions uniformly distributed (reducing the gaps or blank spaces). This function (graphically shown in Fig. 5.25 is applied in a quadratic way, meaning that the smaller the gap within a rule, the better the rule is. Thus, this function calculates the width of the blank spaces with regard to the width of the interval represented by the item i-th, and this relation is calculated for the set of n items included in the pattern that represents the association rule r.

$$F_1 = x^2 = \left(1 - \prod_{i=1}^{n} \frac{BlankWidth_i}{IntervalWidth_i}\right)^2 \tag{5.7}$$

As for the function F_2 (see Eq. (5.8)), the aim is the looking for frequent association rules, so the higher the support value, the better the rule. Nevertheless, it should be noted that maximum support values imply misleading rules as stated by the lift, leverage and conviction measures (see Chap. 2). Hence, this function is defined in such a way that rules having either low support values or extremely high support values are meaningless. In many proposals for mining frequent association rules, support values lower than 50 % are not of interest [15], and the higher the value, the better it is. Therefore, it is considered a function (see Fig. 5.26) that reaches the maximum function value with a support close to 50 % and decreases the function value when the support is close to the maximum (rules that satisfy most of the transactions are not interesting as described in Chap. 2). The mathematical expression of the above representation of the F_2 (see Eq. (5.8)) function is a convex parabola throughout the domain, so using the general equation $(ax^2 + bx + c)$ of the parabola, the constant a should be less than 0. Additionally, it was interesting that the parabola cuts the horizontal axis in the values 0.3 and 1 (the desired range of support values). Thus, the function obtains maximum values for support values in the range $[0.5, 0.8]$.

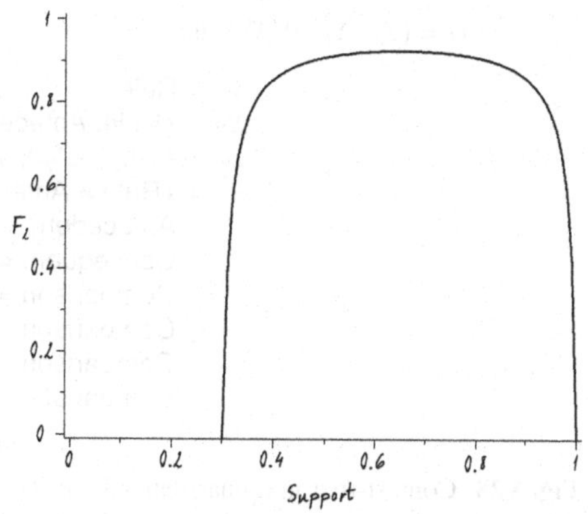

Fig. 5.26 Representation of the F_2 function, which denotes the importance of the support metric

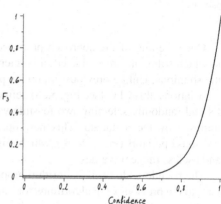

Fig. 5.27 Representation of the F_3 function, which denotes the importance of the confidence metric

$$F_2 = \frac{x(13 - 10x) - 3}{x(13 - 10x) - 2.9} = \frac{Support \times (13 - 10 \times Support) - 3}{Support \times (13 - 10 \times Support) - 2.9} \qquad (5.8)$$

Finally, the third function included in the fitness function is related to the confidence measure, which is one of the most important metrics to determine the reliability of an association rule. As described in Chap. 2, the higher the confidence value of the rule, the more accurate the rule is. Generally speaking, rules with low confidence values are not of interest to the user. In this regard, function F_3 (see Eq. (5.9)) denotes that low confidence values imply low function rates, and these values get higher and higher with the increment of confidence values. As graphically shown in Fig. 5.27, this function provides a low value for those rules having a confidence value lower than 0.8 so really accurate associations are desired.

$$F_3 = x^{10} = confidence^{10} \qquad (5.9)$$

The proposed fitness function $fitness(r) = F_1 + F_2 \times F_3$ obtains the value for an association rule r and this value is in the range $[0, 2]$. In situations where rules comprising only discrete items are discovered, this fitness function discards the function F_1 by using the unity value.

$G = (\Sigma_N, \Sigma_T, P, S)$ with:

S = Rule
Σ_N = {Rule, Antecedent, Consequent, Comparison }
Σ_T = {'∧', '=', '≠', 'IN', 'name', 'value'}
P = {Rule = Antecedent, Consequent ;
          Antecedent = Comparison ;
          Consequent = Comparison ;
          Comparison = '∧', Comparison, Comparison ;
          Comparison = '≠', 'name', 'value' ;
          Comparison = '=', 'name', 'value' ;
          Comparison = 'IN', 'name', 'value', 'value' ;}

Fig. 5.28 Context-free grammar defined for the proposed algorithm that optimizes continuous patterns

Once the aim of the approach proposed in [24] is described, and also the way in which solutions are evaluated, it is interesting to define the CFG used to encode the solutions. Unlike previous grammars proposed in G3P for mining patterns and associations, this CFG (see Fig. 5.28) defines continuous items by using the operator IN and randomly selecting two feasible values within the feasible range of values imposed by the metadata. This new operator is more comprehensible than the previous operators ($<, \geq, >, \leq$), which might provide the user with extremely large and useless range of values.

The pseudo-code of the algorithm proposed in [24] is shown in Fig. 5.29. Similarly to the previous G3P algorithms for mining association rules, this approach uses an elite population to keep the best solutions found along the evolutionary process. Nevertheless, this algorithm includes some features that were not considered in other algorithms like G3PARM [22]. For example, a major feature of this algorithm is related to stopping criterion (see line 11, Fig. 5.29), which is based on the average fitness value of the elite population, so no maximum number of generations is required. Another major feature of this algorithm is its ability to self-adapt the probability value used to apply the genetic operator. Instead of fixing a predefined value, it is modified along the evolutionary process depending on the average fitness value obtained from the elite population (see lines 19 and 20, Fig. 5.29).

5.3.4   Mining Patterns from Relational Databases

Finally, once different G3P algorithms for mining frequent/infrequent patterns, positive/negative patterns and discrete/continuous patterns have been described, it is interesting to analyse a G3P algorithm designed for mining association rules in relational databases [25]. Noted that most approaches for the extraction of patterns of interest and, specially, association rules, look for results in traditional datasets in the form of a single table. However, with the growing interest in the storage

Require: *population_size* {Number of individuals to be considered}
Ensure: *elite_population*
 1: *population* ← ∅
 2: *elite_population* ← ∅
 3: *parents* ← ∅
 4: *offspring* ← ∅
 5: *average_fitness* ← 0
 6: *population* ← generate(*population_size*)
 7: **for** ∀*member* ∈ *population* **do**
 8:    evaluate(*member*) by means of F_1, F_2 and F_3
 9: **end for**
10: **while** *average_fitness* improves **do**
11:    *parents* ← select(*population*)
12:    *offspring* ← apply mutation operator over the set *parents*
13:    **for** ∀*member* ∈ *offspring* **do**
14:       evaluate(*member*) by means of F_1, F_2 and F_3
15:    **end for**
16:    *population* ← update(*population, offspring, elite_population*)
17:    *elite_population* ← best individuals from {*population, offspring, elite_population*}
18:    *average_fitness* ← calculate the average fitness value from *elite_population*
19:    update the mutation probability based on *average_fitness*
20: **end while**
21: return *elite_population*

Fig. 5.29 Pseudo-code of the algorithm proposed in [24] for optimizing continuous patterns

of information, relational databases comprising a series of relations (tables) and relationships have become essential, and this type of datasets is able to represent more complex structures and to store more information than traditional datasets.

Existing ARM proposals for mining association rules cannot directly be applied to relational databases, and the relational structures need to be transformed by joining all the relations into a single relation [29]. Despite the fact that this transformation enables traditional algorithms to be run, the transformation process is not an easy task and suffers from both a high computational time and a deviation in the support quality measure (each transaction in a relation could be duplicated as the result of a join, so the same item could be read and stored multiple times). Support deviation has been studied by many researchers [8, 13], and the support quality measure is defined as the number of transactions in the result of a join of the relations in the database. In this definition, it is crucial to clarify that the support of a pattern strongly depends on how well its items are connected. For instance, having the relations *Customers* and *Items* in a market basket relational database (see Fig. 5.30), the city *Cambridge* appears in the relation *Customers* with a probability of 0.50, i.e. satisfies 50 % of the transactions. On the other hand, the same city appears in 67 % of the transactions in the result of a join (see Fig. 5.30c). Therefore, joining different relations into a single relation could introduce distortions in their support values so counting rows to calculate the support measure is not correct— each row does not identify a single customer.

a

Customer	Name	City
1	Jack	Birmingham
2	Grace	Cambridge
3	Grace	Bradford
4	Jack	Cambridge

b

Customer	Items	Invoice
1	Banana, Bread, Butter, Orange	6.5
2	Banana, Bread, Milk	5.4
2	Bread, Butter, Milk	6.1
3	Banana, Butter, Milk	5.8
4	Bread, Butter, Milk, Orange	6.6
4	Banana, Butter, Milk, Orange	7.4

c

Customer	Name	City	Items	Invoice
1	Jack	Birmingham	Banana, Bread, Butter, Orange	6.5
2	Grace	Cambridge	Banana, Bread, Milk	5.4
2	Grace	Cambridge	Bread, Butter, Milk	6.1
3	Grace	Bradford	Banana, Butter, Milk	5.8
4	Jack	Cambridge	Bread, Butter, Milk, Orange	6.6
4	Jack	Cambridge	Banana, Butter, Milk, Orange	7.4

Fig. 5.30 Sample market basket comprising two relations and the result of joining both relations. (**a**) Customers relation; (**b**) Items relation; (**c**) Customers ⋈ Items

In 2002, Jiménez et al. [14] proposed a way of representing relational databases by means of trees, describing two different representation schemes: key-based and object-based. As described by the authors, each of these representations starts from a particular relation, namely the target relation, which is selected by the end user according to the user's specific goals. The key-based tree representation is based on the fact that each transaction is identified by its primary key. Therefore, a root node is represented by its primary key and the children of this primary key will be the remaining attribute values from the transaction in the relation. On the contrary, in the object-based tree, the root represents a transaction, and the children represent all the attribute values, including the primary key.

Based on the previous idea of representing relational databases as sets of trees, Luna et al. [25] proposed the use of G3P for mining association rules in relational environments. To this end, they proposed a new CFG (see Fig. 5.31) to represent different associations in the form of tree structures. Once the tree structure is defined according to the CFG, the next step is to assign values to the leaf nodes depending on the specific relational database used. Noted that the target relation or target table is used as a root table from which any path to the remaining tables starts [14], and this first table is randomly selected from the set of tables in the relational database. According to the values included in each condition, they are related to the table in which they were defined. Finally, a table name is randomly assigned to the leaf node '*Table*' selected from all tables related to this target relation.

$G = (\Sigma_N, \Sigma_T, P, S)$ with:

S = Rule
Σ_N = {Rule, Antecedent, Consequent, Condition }
Σ_T = {'∧', '*TargetTable*', '*Table*', '=', '≠', '*IN*', '*name*', '*value*'}
P = {Rule = '*TargetTable*', Antecedent, Consequent ;
         Antecedent = Condition | '∧', Condition, Antecedent |
                        '∧', Condition, '*Table*', Antecedent ;
         Consequent = Condition | '∧', Condition, Consequent |
                        '∧', Condition, '*Table*', Consequent ;
         Condition = '≠', '*name*', '*value*' ;
         Condition = '=', '*name*', '*value*' ;
         Condition = '*IN*', '*name*', '*value*', '*value*' ;}

Fig. 5.31 Context-free grammar defined for representing association rules in relational environments

Fig. 5.32 Sample individual encoded by the proposed algorithm for mining association rules in relational environments

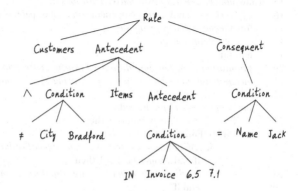

In order to illustrate the structure of a sample individual encoded by means of the proposed CFG, let us consider again the sample database shown in Fig. 5.30, where there is a $1 : N$ relationship between tables *Customers* and *Items*. From this database, a sample individual encoded by the CFG is shown in Fig. 5.32, which describe the association rule: *from all the customers in the database, if their city is not Bradford and they have a market basket whose invoice is in the range [6.5, 7.1], then they should be named Jack.*

As for the genetic operators used in this G3P algorithm for mining associations in relational databases, the operators are the same as those described by G3PARM [22], which obtained interesting results. Thus, the main idea is to replace the least frequent item defined within a parent to produce a new one that appear more frequently in the database. Additionally, the schema of this algorithm follows the general evolutionary schema with an elite population (see Fig. 5.33). The main feature of this algorithm is its ability to create valid individuals in the first generation, so this algorithm guarantees that each individual covers at least one transaction. If an individual generated conformant to the CFG does not satisfy any transaction, then a new random individual is generated (see lines 7–13, Fig. 5.33). The process continues until the maximum population size is reached.

Require: *population_size, maxGenerations* {Number of individuals and generations considered}
Ensure: *elite_population*
 1: *population* ← ∅
 2: *elite_population* ← ∅
 3: *offspring* ← ∅
 4: *number_generations* ← 0
 5: **while** size of *population* smaller than *population_size* **do**
 6:     create *member*
 7:     evaluate(*member*)
 8:     **if** getSupport(*member*) > 0 **then**
 9:         *elite_population* ← (*elite_population* ∪ *member*)
10:     **end if**
11: **end while**
12: **for** ∀*member* ∈ *population* **do**
13:     evaluate(*member*)
14: **end for**
15: **while** *number_generations* < *maxGenerations* **do**
16:     *offspring* ← apply genetic operators on a set of parents selected from *population*
17:     **for** ∀*member* ∈ *offspring* **do**
18:         evaluate(*member*)
19:     **end for**
20:     *population* ← update(*population, offspring, elite_population*)
21:     *aux_population* ← *population* ∪ *offspring* ∪ *elite_population*
22:     *elite_population* ← ∅
23:     **for** ∀*member* ∈ *population* **do**
24:         **if** *member* ∉ *elite_population* **then**
25:             **if** getFitness(*member*) > 0 **then**
26:                 **if** getConfidence(*member*) > *confidenceTreshold* **then**
27:                     **if** getLift(*member*) > 1 **then**
28:                         *elite_population* ← (*elite_population* ∪ *member*)
29:                     **end if**
30:                 **end if**
31:             **end if**
32:         **end if**
33:     **end for**
34:     *number_generations* + +
35: **end while**
36: return *elite_population*

Fig. 5.33 Pseudo-code of the G3P algorithm for mining association rules in relational databases

5.4 Successful Applications

One of the main features of G3P is its ability to introduce subjective knowledge into
the pattern mining process, enabling the structure of the pattern or the form of the
association rule to be predefined by the end user according to his/her preferences.
This ability to introduce subjectivity is really useful in many application domains
and, specially, in educational tasks [34]. In this specific domain, the subjectivity
plays a really important role since the knowledge of the instructors is really
important. Additionally, the educational application domain requires, sometimes,

the extraction of comprehensible knowledge since it is then analysed by users with no background in data mining. Hence, it demonstrates the usefulness of applying G3P approaches to educational domains.

A really interesting G3P approach to provide feedback to instructors from multiple-choice quizzes was proposed in [35], which enables quizzes and courses to be improved. The aim of applying G3P to this field was to discover interesting relationships to aid the instructor in decision making about how to improve both the quiz and the corresponding course that contains the concepts evaluated by the quiz. To evaluate the usefulness of the extracted knowledge, authors analysed the rules discovered and applied them into the course to evaluate the improvements. To evaluate these improvements, different groups of students enrolled in the course were analysed before and after applying the extracted knowledge in the form of association rules, and this analysis was carried out by comparing the final score obtained by the students. The experimental analysis when comparing the group of students showed that the updates carried out to improve the courses/quizzes had a positive impact on student learning.

The extraction of infrequent or abnormal behaviour by means of G3P has also been of interest in educational environments [26]. Noted that infrequent associations might allow the instructor to verify a set of rules concerning certain unusual learning problems, for instance dealing with students with special needs. Thus, this information could help the instructor to discover a minority of students who may need specific support in their learning process. The idea is to find infrequent relations that discover unusual behaviours in the form of **IF** *a student spends high time doing task* **THEN** *the student fails the course*. This could be an abnormal behaviour that should be analysed to determine why they do not achieve the aim and which specific needs they require.

Finally, G3P has also been successfully applied to the extraction of useful knowledge from Moodle learning management system. Noted that data in Moodle is saved as a relational database, so the extraction of patterns of interest is not a trivial task. To solve it, a really interesting G3P approach for mining patterns in relational environments were designed [25]. The approach was applied to data gathered from Moodle from a Spanish university, and the results were really promising, describing the behaviuor of students in a specific course. For instance, one of the rules discovered described the level of dedication of students based on their expectation at the beginning of the course. As a matter of example, a really interesting rule described that 98 % of the students that did not have high expectations to pass the course, then they will dedicate enough time to the course. This description is quite interesting since determines that the expectations present a high relation to the time spent in the course.

References

1. C. C. Aggarwal and J. Han. *Frequent Pattern Mining*. Springer International Publishing, 2014.
2. W. Banzhaf, F. D. Francone, R. E. Keller, and P. Nordin. *Genetic programming: an introduction. On the automatic evolution of computer programs and its applications*. Morgan Kaufmann Publishers Inc., San Francisco, CA, USA, 1998.
3. M. J. del Jesús, J. A. Gámez, P. González, and J. M. Puerta. On the discovery of association rules by means of evolutionary algorithms. *Wiley Interdisciplinary Reviews: Data Mining and Knowledge Discovery*, 1(5):397–415, 2011.
4. A. A. Freitas. *Data Mining and Knowledge Discovery with Evolutionary Algorithms*. Springer-Verlag Berlin Heidelberg, 2002.
5. M. Fuchs. Crossover versus mutation: an empirical and theoretical case study. In *In Proceedings of the third annual conference on Genetic Programming*, GP '98, pages 78–85, 1998.
6. M. Gendreau and J. Potvin. *Handbook of Metaheuristics*. Springer Publishing Company, Incorporated, 2nd edition, 2010.
7. B. Goethals. Survey on Frequent Pattern Mining. Technical report, Technical report, HIIT Basic Research Unit, Department of Computer Science, University of Helsinki, Finland, 2003.
8. B. Goethals, W. Le Page, and M. Mampaey. Mining interesting sets and rules in relational databases. In *Proceedings of the ACM Symposium on Applied Computing*, pages 997–1001, Sierre, Switzerland, March 2010.
9. B. Goethals, S. Moens, and J. Vreeken. MIME: A Framework for Interactive Visual Pattern Mining. In D. Gunopulos, T. Hofmann, D. Malerba, and M. Vazirgiannis, editors, *Machine Learning and Knowledge Discovery in Databases*, volume 6913 of *Lecture Notes in Computer Science*, pages 634–637. Springer Berlin Heidelberg, 2011.
10. P. González-Espejo, S. Ventura, and F. Herrera. A Survey on the Application of Genetic Programming to Classification. *IEEE Transactions on Systems, Man and Cybernetics: Part C*, 40(2):121–144, 2010.
11. M. Gorawski and P. Jureczek. Extensions for Continuous Pattern Mining. In *Proceedings of the 2011 International Conference on Intelligent Data Engineering and Automated Learning*, IDEAL 2011, pages 194–203, Norwich, UK, 2011.
12. J. H. Holland. *Adaptation in Natural and Artificial Systems*. The University of Michigan Press, 1975.
13. V. C. Jensen and N. Soporkar. Frequent itemset counting across multiple tables. In *Proceedings of the 4th Pacific-Asia Conference on Knowledge Discovery and Data Mining*, PADKK '00, pages 49–61, Kyoto, Japan, April 2000.
14. A. Jiménez, F. Berzal, and J. C. Cubero. Using trees to mine multirelational databases. *Data Mining and Knowledge Discovery*, 24(1):1–39, 2012.
15. C. S. Kanimozhi and A. Tamilarasi. An automated association rule mining technique with cumulative support thresholds. *International Journal Open Problems in Computer Science and Mathematics*, 2(3):427–438, 2009.
16. Y. S. Koh and N. Rountree. Finding sporadic rules using apriori-inverse. In *Proceedings of the 9th Pacific-Asia Conference on Advances in Knowledge Discovery and Data Mining*, PAKDD'05, pages 97–106, Hanoi, Vietnam, 2005.
17. Y. S. Koh and N. Rountree. *Rare Association Rule Mining and Knowledge Discovery: Technologies for Infrequent and Critical Event Detection*. Information Science Reference, Hershey, New York, 2010.
18. J. R. Koza. *Genetic Programming: On the Programming of Computers by Means of Natural Selection (Complex Adaptive Systems)*. A Bradford Book, 1 edition, 1992.
19. T. Li and X. Li. Novel alarm correlation analysis system based on association rules mining in telecommunication networks. *Information Sciences*, 180(16):2960–2978, 2010.
20. S. Luke and L. Spector. A comparison of crossover and mutation in genetic programming. In *Proceedings of the Second Annual Conference on Genetic Programmin*, GP '97, pages 240–248. Morgan Kaufmann, 1997.

21. J. M. Luna, J. R. Romero, and S. Ventura. G3PARM: A Grammar Guided Genetic Programming Algorithm for Mining Association Rules. In *Proceedings of the IEEE Congress on Evolutionary Computation*, IEEE CEC 2010, pages 2586–2593, Barcelona, Spain, 2010.

22. J. M. Luna, J. R. Romero, and S. Ventura. Design and behavior study of a grammar-guided genetic programming algorithm for mining association rules. *Knowledge and Information Systems*, 32(1):53–76, 2012.

23. J. M. Luna, J. R. Romero, and S. Ventura. On the adaptability of G3PARM to the extraction of rare association rules. *Knowledge and Information Systems*, 38(2):391–418, 2014.

24. J. M. Luna, J. R. Romero, C. Romero, and S. Ventura. Reducing gaps in quantitative association rules: a genetic programming free-parameter algorithm. *Integrated Computer Aided Engineering*, 21(4):321–337, 2014.

25. J. M. Luna, A. Cano, and S. Ventura. Genetic programming for mining association rules in relational database environments. In A. H. Gandomi, A. H. Alavi, and C. Ryan, editors, *Handbook of Genetic Programming Applications*, pages 431–450. Springer International Publishing, 2015.

26. J. M. Luna, C. Romero, J. R. Romero, and S. Ventura. An Evolutionary Algorithm for the Discovery of Rare Class Association Rules in Learning Management Systems. *Applied Intelligence*, 42(3):501–513, 2015.

27. J. Mata, J. L. Alvarez, and J. C. Riquelme. Mining numeric association rules with genetic algorithms. In *Proceedings of the 5th International Conference on Artificial Neural Networks and Genetic Algorithms*, ICANNGA 2001, pages 264–267, Taipei, Taiwan, 2001.

28. R. McKay, N. Hoai, P. Whigham, Y. Shan, and M. O'Neill. Grammar-based Genetic Programming: a Survey. *Genetic Programming and Evolvable Machines*, 11:365–396, 2010.

29. E. K. K. Ng, A. W. Fu, and K. Wang. Mining association rules from stars. In *In Proceedings of the 2002 IEEE International Conference on Data Mining*, ICDM 2002, pages 322–329, Maebashi City, Japan, 2002.

30. N. Ordoñez, C. Ezquerra and C. Santana. Constraining and Summarizing Association Rules in Medical Data. *Knowledge and Information Systems*, 9, 2006.

31. R. Poli, W. B. Langdon, and N. F. McPhee. *A Field Guide to Genetic Programming*. Lulu Enterprises, UK Ltd, 2008.

32. A. Rahman, C. I. Ezeife, and A. K. Aggarwal. Wifi miner: An online apriori-infrequent based wireless intrusion system. In *Proceedings of the 2nd International Workshop in Knowledge Discovery from Sensor Data*, Sensor-KDD '08, pages 76–93, Las Vegas, USA, 2008.

33. C. Romero and S. Ventura. Educational data mining: a review of the state of the art. *IEEE Transactions on Systems, Man, and Cybernetics, Part C*, 40(6):601–618, 2010.

34. C. Romero, S. Ventura, M. Pechenizkiy, and R. Baker. *Handbook of Educational Data Mining*. Data Mining and Knowledge Discovery Series. CRC Press, 2010.

35. C. Romero, A. Zafra, J. M. Luna, and S. Ventura. Association rule mining using genetic programming to provide feedback to instructors from multiple-choice quiz data. *Expert Systems*, 30(2):162–172, 2013.

36. D. Sánchez, J. M. Serrano, L. Cerda, and M. A. Vila. Association Rules Applied to Credit Card Fraud Detection. *Expert systems with applications*, (36):3630–3640, 2008.

37. P. A. Whigham. Grammatically-based genetic programming. In *Proceedings of the Workshop on Genetic Programming: From Theory to Real-World Applications*, pages 33–41, Tahoe City, California, USA, 1995.

38. M. L. Wong and K. S. Leung. *Data Mining Using Grammar-Based Genetic Programming and Applications*. Kluwer Academic Publishers, Norwell, MA, USA, 2000.

39. X. Yan, C. Zhang, and S. Zhang. Genetic algorithm-based strategy for identifying association rules without specifying actual minimum support. *Expert Systems with Appications*, 36:3066 – 3076, 2009.

Chapter 6
Multiobjective Approaches in Pattern Mining

Abstract In pattern mining, solutions tend to be evaluated according to several conflicting quality measures and, sometimes, these measures have to be optimized simultaneously. The problem of optimizing more than one objective function is known as multiobjective optimization, which together with evolutionary computation have given rise to evolutionary multiobjective optimization. This issue is well addressed in this chapter, which describes the usefulness of evolutionary multiobjective optimization in the pattern mining field. First, some fundamental concepts of evolutionary multiobjective optimization are introduced, including metrics to evaluate the quality of the solutions, and measures to be combined for the pattern mining problem. Then, different multiobjective approaches in the pattern mining field are described, which are divided into genetic algorithms, genetic programming approaches and other kind of algorithms like evolutionary computation and swarm intelligence. Finally, some useful real-world applications are analysed.

6.1 Introduction

In previous chapters, numerous proposals for mining patterns of interest from data have been proposed [2, 37]. Some of these proposals are constraint-based and they mine every association between patterns satisfying a series of constraints, such as minimum quality thresholds for different metrics [13]. These proposals extract the set of solutions that are optimal according to some interestingness measures [10, 15, 34], and these proposals are considered as particularly useful in domains where the set of rules discovered is too large to be understood by the end user.

The process of extracting patterns and association between patterns from a dataset often requires the application of more than one quality measure [14]. In many situations, such quality measures involve conflicting objectives so the improvement of one objective may lead to deterioration of another. In this situation, it does not exist a single solution that optimizes all objectives simultaneously. Instead, the best trade-off solutions, known as pareto optimal front, will form the final solution to the problem and no one of these pareto solutions can be said to be better than the others.

© Springer International Publishing Switzerland 2016

S. Ventura, J.M. Luna, *Pattern Mining with Evolutionary Algorithms*,

DOI 10.1007/978-3-319-33858-3_6

According to some authors [8], classical optimization methods (including the multicriterion decision-making methods) have been used by transforming multi-objective into single-objective optimization problems. To do so, these methods consider just one particular solution from the pareto front at a time, so they are required to be applied multiple times for finding different solutions. To overcome this drawback, many different multiobjective evolutionary algorithms (MOEAs) have been proposed [7, 9] since they are able to find multiple pareto optimal solutions in one single run. As described in Chap. 3, evolutionary algorithms usually work with a population as a set of solutions, and this set of solutions can be considered as a pareto front that moves toward the optimal region.

As it was demonstrated by many authors, the use of evolutionary algorithms for mining patterns of interest plays an important role [19, 31]. This issue, together with the necessity of optimizing not only a single quality measure but also more than one, has given rise to the use of MOEAs in pattern mining [6]. Thus, any pattern mining algorithm based on MOEAs has two different goals: (a) to discover a set of patterns that forms a good pareto optimal front, and (b) to extract a set of patterns as diverse as possible according to the objective quality measures.

In the pattern mining field, different authors have considered the problem as a multiobjective optimization one, where individuals are encoded by means of a linear chromosome for representing patterns defined in both continuous and discrete domains [3, 11]. Additionally, the use of both positive and negative patterns has been addressed by different researchers [21]. Other authors have considered the importance of introducing syntax constraints into MOEA for mining patterns [20]. These syntax constraints can be modelled by means of grammars, which can be adapted to the problem under study, considering different structures of the patterns, types of items and logical operators. Finally, other multiobjective approaches based on swarm intelligence have been considered for mining patterns of interest [25, 26].

Since the pattern mining task requires more than a single quality measure to be optimized, the use of multiobjective optimization in pattern mining has been considered by different researchers in different application domains. One of these application domains is molecular biology, where one of the main goals is the reconstruction of gene regulatory processes by the inference of regulatory interactions among genes. In this regard, the use of pattern mining helps to infer the relationships between genes from an organism in a particular biological process [23].

6.2 General Issues

As previously described, the use of multiobjective evolutionary algorithms (MOEAs) for mining patterns of interest [3, 11] enables more than a single quality measure to be optimized. These algorithms optimize all objectives simultaneously, discovering the best trade-off solutions, and none of these solutions can be said to be better than the others. In this section, the multi-objective optimization problem is properly described, denoting some quality measures to determine the interest of the solutions.

6.2.1 Multiobjective Optimization

The multiobjective optimization problem (also known by some authors as multicriteria optimization, multiperformance or vector optimization problem) can be defined [5] as:

> A vector of decision variables which satisfies constraints and optimizes a vector function whose elements represent the objective functions. These functions form a mathematical description of performance criteria which are usually in conflict with each other. Hence, the term "optimize" means finding such a solution which would give the values of all the objective functions acceptable to the decision maker.

In an optimization problem, decision variables are defined in a numerical domain in which values are selected. These variables are represented as $\{x_j : j = 1, 2, \ldots, n\}$, and the set x of decision variables is represented as $x = [x_1, x_2, \ldots, x_n]^T$, where T denotes the transposition of the column vector to the row vector. Additionally, the decision variables must satisfy some constraints or restrictions in order to consider a certain solution acceptable. All these constraints might be expressed in the form of mathematical inequalities $\{g_i(x) \leq 0 : i = 1, \ldots, m\}$ or equalities $\{h_j(x) = 0 : j = 1, \ldots, p\}$. Thus, a multiobjective optimization problem can be defined as a minimization or a maximization problem, i.e. $minimize[f_1(x), f_2(x), \ldots, f_k(x)]$ or $maximize[f_1(x), f_2(x), \ldots, f_k(x)]$ where k is the number of different objective functions $\mathscr{F} = \{f_1, \ldots, f_k\}$, and the aim is to determine the set of values $x_1^*, x_2^*, \ldots, x_n^*$ that satisfy the constraints and which yield the optimum values for all the objective functions $f \in \mathscr{F}$.

Rarely occurs that a single point is able to simultaneously optimize all the objective functions of a multiobjective optimization problem. Therefore, the optimization problem is considered as the looking for a set of solutions with good trade-offs rather than a single solution. The set of solutions with good trade-offs is known as pareto optimal front [32]. A vector of decision variables x is pareto optimal if there is no other vector x' that dominates x. If the aim of the problem is to minimize the objective functions, then it is said that x' dominates x if and only if $f_i(x') \leq f_i(x)$ for all $i = 1, \ldots, k$ and $f(x') < f(x)$ for at least one $f \in \mathscr{F}$. In such situations where the aim is not to minimize but to maximize the objective functions, then it is said that a vector of decision variables x belongs to the pareto optimal front if there is no other x', where $f_i(x') \geq f_i(x)$ for all $i = 1, \ldots, k$ and $f_i(x') > f_i(x)$ for at least one $f \in \mathscr{F}$. In both cases, the vector x of decision variables is known as a solution to the optimization problem.

As depicted in Fig. 6.1, having ten different solutions (s_1, s_2, ..., s_{10}) and two objectives (f_1 and f_2) to be maximized, the pareto optimal front comprises the

Fig. 6.1 Representation of sample pareto optimal front that comprises the following solutions: s_7, s_8, s_9 and s_{10}

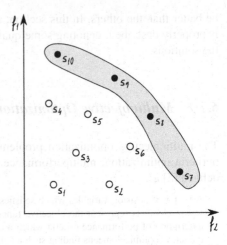

solutions s_7, s_8, s_9 and s_{10}. These solutions are equally acceptable because neither of them is better than the remaining solutions within this set for all the objectives. Analysing solution s_7, it belongs to the pareto optimal front since it obtains the highest value for the function f_2 so there is no other solution s_i such as $f_2(s_i) > f_2(s_7)$. Additionally, s_8 also belongs to the pareto front since there is no other solution that dominates it. On the contrary, s_6 does not belong to the pareto optimal front, since $f_1(s_8) > f_1(s_6)$ and $f_2(s_8) > f_2(s_6)$, i.e., there exists at least one solution (s_8) that dominates s_6

6.2.2   Quality Indicators of the Pareto Front

In a multiobjective optimization problem [5], the validation of the interest of the results requires a precise analysis since there is a single solution but a set of them equally promising. Thus, the qualitative assessment of the results in any multiobjective optimization problem becomes hard with the increasing number of solutions in the pareto optimal front.

Some authors [38] have considered that any metric defined to evaluate the quality of the pareto front should satisfy the following issues: (1) minimize the distance of the obtained pareto front with respect to the global or true pareto front (assuming the location is previously known); (2) maximize the spread of solutions found; (3) maximize the number of elements of the pareto optimal front. Considering the previous issues, different researchers have described a variety of metrics to evaluate the quality of the pareto optimal fronts [1].

The error ratio (ER) metric was proposed by Van Veldhuizen [35] as a way to determine the percentage of solutions that are not members of the true pareto optimal front. Thus, this metric is applicable just in those cases where the true pareto front is known. ER is mathematically defined as shown in Eq. (6.1), where n is the

Fig. 6.2 Representation of both the obtained pareto front (*black dots*) and the true pareto front (*white dots*) for a optimization problem where the aim is to maximize

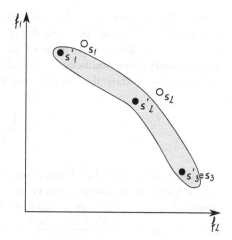

number of solutions in the current set, $s_i = 0$ if solution i-th is a member of the pareto front, and $s_i = 1$ otherwise. Thus, it should be noted that $ER = 0$ indicates an ideal behaviour, describing that all solutions belong to the pareto optimal front.

$$ER = \frac{\sum_{i=1}^{n} s_i}{n} \tag{6.1}$$

Analysing Fig. 6.2, the number of solutions in the obtained pareto front is $n = 3$, and there is one of these solutions that does not belong to the true pareto front. Hence, the value for error ratio metric for this example is $ER = \frac{2}{3} = 0.667$.

Generational distance (GD) is another metric for estimating how far, in average, the solutions in the obtained pareto front are from those in the true pareto optimal front [36]. Thus, similarly to the ER metric, the GD metric (see Eq. (6.2)) requires that the true pareto front be known beforehand. Note that d_i is the Euclidean distance between each solution, s_i, of the obtained pareto front and the closest solution in the true pareto front to that solution s_i. In such situations where $GD = 0$, then both pareto fronts are the same.

$$GD = \frac{\sqrt{\sum_{i=1}^{n} d_i^2}}{n} \tag{6.2}$$

Considering the solutions shown in Fig. 6.2, and the values $s_1' = (0.1, 0.9)$, $s_1 = (0.15, 0.95)$, $s_2' = (0.7, 0.65)$, $s_2 = (0.75, 0.7)$, and $s_3' = s_3 = (0.9, 0.15)$, we obtain the following distances: $d_1 = \sqrt{0.005}$, $d_2 = \sqrt{0.005}$, and $d_3 = \sqrt{0}$. Hence, $GD = \frac{d_1^2 + d_2^2 + d_3^2}{3} = 0.0033$.

The quality of a pareto optimal front can also be measured according to how well the solutions in this front are distributed. In this regard, the spacing (SP) metric [5] was proposed to determine the distance between solutions in the front (see Eq. (6.3)). It should be noted that $d_i = min_j(|f_1^i(x) - f_1^j(x)| + |f_2^i(x) - f_2^j(x)|)$, $i, j = 1, \ldots, n$, and $\overline{d}$ is the mean of all d_i. When $SP = 0$, then all solutions are spaced evenly apart. Additionally, it should be noted that this metric does not require the researcher to know the true pareto optimal front.

$$SP = \sqrt{\frac{1}{n-1} \sum_{i=1}^{n} (\overline{d} - d_i)^2} \tag{6.3}$$

Again considering the obtained solutions (black dots) shown in Fig. 6.2, and whose values were described above, we obtain the following values for d_i: $d_1 = min(0.85, 1.55) = 0.85$, $d_2 = min(0.85, 0.7) = 0.7$, and $d_3 = min(1.55, 0.7) = 0.7$; whereas $\overline{d} = \dfrac{0.85 + 0.7 + 0.7}{3} = 0.75$. Hence, $\sum_{i=1}^{n} (\overline{d} - d_i)^2 = ((0.75 - 0.85)^2 + (0.75 - 0.7)^2 + (0.75 - 0.7)^2) = 0.015$ and $SP = \sqrt{\dfrac{1}{3-1} 0.015} = 0.0866$

Finally, another interesting quality measure is the hyperarea (HA) [1] obtained for two objective functions, also known as hypervolume (HV) if there are two or more objective functions. This quality measure calculates the area of coverage of the pareto optimal front (see Fig. 6.3) according to the objective considered in the optimization problem. Considering the sample pareto optimal front formed by solutions s_1', s_2' and s_3', the area obtained by s_1' is 0.09, the area formed by s_2' (without considering the area in common with s_1') is 0.39, and the area formed by s_3' (without considering the area in common with s_1' and s_2') is 0.03, giving rise to the value $HV = 0.09 + 0.39 + 0.03 = 0.51$.

The HA quality measure can be also applied to the true pareto front. Thus, it is possible to obtain the ratio or differences between the obtained pareto front and the true pareto front. This relationship between the HA value for both pareto fronts is

Fig. 6.3 Representation of the hypervolume quality measure for a sample pareto optimal front

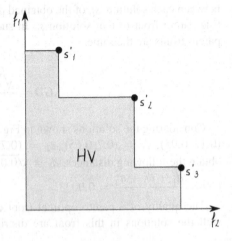

Fig. 6.4 Representation of the hypervolume quality measure for true pareto optimal front

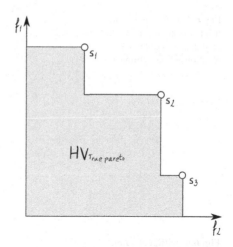

known as hyperarea ratio (HR). Considering the area covered by the true pareto front (see Fig. 6.4), it is calculated as $HA = 0.1425 + 0.42 + 0.0225 = 0.585$. Thus, the value for the HR metric is equal to $HR = \dfrac{0.51}{0.585} = 0.8718$, so the obtained pareto front is close to the optimum.

6.2.3 Quality Measures to Optimize in Pattern Mining

In the pattern mining field, different authors have considered the task of mining patterns of interest as a multiobjective optimization problem where the objective functions are some of the quality measures described in Chap. 2. Support and confidence are broadly conceived as the finest quality measures in the pattern mining field and, consequently, a great variety of proposals make use of them [30]. Support calculates the percentage of transactions satisfied by a specific pattern, whereas confidence quality measure determines the reliability of implication of the rule, so the higher its value, the more accurate the rule is.

Support and confidence are broadly conceived as the finest quality measures in the pattern mining field and, consequently, a great variety of proposals make use of them [30]. In this regard, both support and confidence quality measures have been proposed by some researchers [20] as objective functions to be considered in a multiobjective optimization problem. Nevertheless, these two quality measures might not be appropriate to this end since the increase of the support values also imply the increase of the confidence values. It should be noted that these two metrics are related as shown in Fig. 6.5, illustrating the feasible area in which any solution can be located (confidence is always greater or equal to support as demonstrated in Chap. 2). As shown, despite the fact that *Pareto*1 is dominated by the other pareto fronts, this pareto front includes a good set of solutions since they are equally spaced

Fig. 6.5 Different pareto
fronts for support and
confidence as objective
functions to be optimized

Fig. 6.6 Different pareto
fronts for support and lift as
objective functions to be
optimized

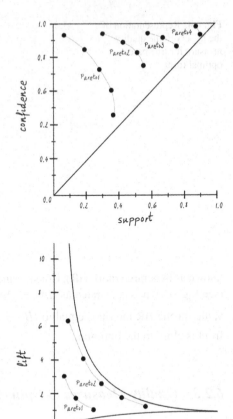

and comprise huge range of values for both support and confidence measures. On
the contrary, *Pareto*4 dominates the others, but it includes solutions whose values
tend to be similar, so these support-confidence measures are not really appropriate
to be used in a multiobjective optimization problem.

Since confidence measure should not be considered to be optimized together
with support measure, some researchers [20] have considered support and lift as
good quality measures to be optimized at time. Lift measure was proposed as a
good metric to quantify the interest of associations between patterns [4]. One of the
main characteristics of this quality measure is that it satisfies the three properties
provided by Piatetsky-Shapiro [27]. Let us consider now two sample pareto fronts
(see Fig. 6.6) for support-lift metrics. As shown, *Pareto*1 is dominated by *Pareto*2,
which includes a set of solutions equally spaced and with a huge range of values for
both support and lift. Thus, the behaviour is completely different to the one obtained
when support and confidence were selected as objective functions to be optimized.

All of this lead to the conclusion that the choice of a good pair of objective functions in a multiobjective optimization problem is a cornerstone. Thus, it is really important to analyse all existing quality measures in the pattern mining field (see Chap. 2) to be able to obtain fine solutions, i.e. a promising pareto optimal front.

6.3 Algorithmic Approaches

As described in previous sections, different authors have considered the pattern mining task as a multiobjective optimization problem [28], and different MOEAs have been proposed [7, 9] since they are able to find multiple pareto optimal solutions in one single run. In this section, different MOEAs proposed by different researchers are described. First, some algorithms that encode solutions by means of linear chromosomes for representing patterns are analysed. Then, it is also described the importance of introducing syntax constraints into MOEAs for mining patterns. Finally, other multiobjective approaches based on evolutionary computation and swarm intelligence have been considered.

6.3.1 Genetic Algorithms

A really interesting multiobjective approach for mining patterns of interest was proposed by Ghosh et al. [11]. Authors proposed an algorithm for mining patterns and association between patterns in discrete domains. To this end, the algorithm encodes rules as linear chromosomes by using fixed-length strings of genes where the i-th gene denotes the i-th item in the database. Additionally, each item includes two extra tag bits. If the two tag bits are 00 then the item represented by these two bits will appear in the antecedent of the rule, whereas the value 11 describes that the item will appear in the consequent of the rule. Any of the other two combinations (01 or 10) will indicate the absence of the item in the rule. For example, Fig. 6.7 shows a sample individual that represents the rule $C \wedge D \rightarrow B$. We assumed that only four items exist in the dataset.

As in a multi-objective optimization algorithm, the aim of the algorithm (see Fig. 6.8) is to look for a set of good solutions. In order to reach this goal, the best solutions found over generations are included into a separate set of solutions. This idea is quite similar to one of *elitism* used by genetic algorithms, which keeps the

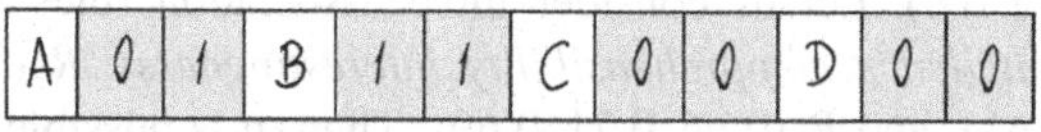

A	0	1	B	1	1	C	0	0	D	0	0

Fig. 6.7 Example of an individual that represents the rule $C \wedge D \rightarrow B$. Note that the first item does not appear in the rule since they have the value 01 in the two tag bits

Require: *population_size* {Number of individuals and generations considered}
Ensure: *pareto_front*
1: *population* ← ∅
2: *parents* ← ∅
3: *offspring* ← ∅
4: *population* ← generate(*population_size*)
5: *pareto_front* ← ∅
6: **for** ∀*member* ∈ *population* **do**
7:     evaluate(*member*) by means of *confidence*, *comprehensiblity* and *interestingness*
8: **end for**
9: **for** ∀*member* ∈ *population* **do**
10:     compute fitness for *member* based on the dominance property
11: **end for**
12: **while** termination condition is not reached **do**
13:     *parents* ← roulette-wheel selection using the set *population*
14:     *offspring* ← apply genetic operators on the set *parents*
15:     **for** ∀*member* ∈ *offspring* **do**
16:         evaluate(*member*) by means of *confidence*, *comprehensiblity* and *interestingness*
17:     **end for**
18:     **for** ∀*member* ∈ *population* **do**
19:         compute fitness for *member* based on the dominance property
20:     **end for**
21:     *population* ← update(*population,offspring*)
22:     *pareto_front* ← update(*population,offspring*)
23: **end while**

Fig. 6.8 Pseudo-code of the algorithm proposed by Ghosh et al. [11]

best solution/s along the evolutionary process. The goodness of a specific solution will be determined by three quality measures: confidence, comprehensibility and interestingness (see line 8, Fig. 6.8). Whereas the first one is widely known in the pattern mining field, the other two quality measures were proposed to quantify the interest of the associations between patterns [17]. Comprehensibility is a subjective quality measure defined as the number of items involved in the antecedent part of a rule with regard to the consequent part of the rule, i.e. *comprehensibility* = $log(1 + |Y|)/log(1 + |R|)$. Given an association rule R, $|R|$ denotes the number of items within the whole rule, whereas $|Y|$ defines the number of items within the consequent of the rule. Additionally, interestingness is used in pattern mining to quantify how surprising a rule might be to the users. Mathematically, this quality measure is defined as *interestingness*$(X \rightarrow Y) =$ *confidence*$(X \rightarrow Y) \times (support(X \rightarrow Y)/support(Y)) \times (1 - (support(X \rightarrow Y)/n))$, n defined as the number of transactions in the dataset. Once these three quality measures are calculated, solutions are ranked based on the concept of dominance (see line 20, Fig. 6.8). Let us consider three association rules with the following values R_i(*confidence, comprehensibility, interestingness*): $R_1(0.92, 0.59, 0.61)$, $R_2(0.92, 0.71, 0.61)$, and $R_3(0.96, 0.71, 0.69)$. Then, it is possible to assert that R_3 dominates R_2, and R_2 dominates R_1, so the algorithm will assign the fitness values as *fitness*$(R_3) >$ *fitness*(R_2)*fitness*(R_1).

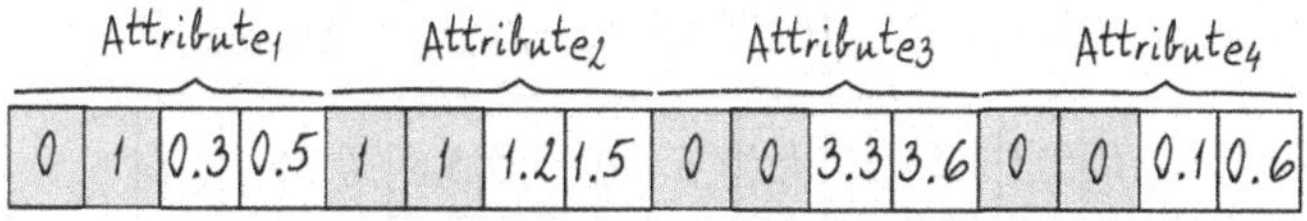

Fig. 6.9 Example of an individual that represents the rule $Att_3[3.3, 3.6] \wedge Att_4[0.1, 0.6] \rightarrow Att_2[1.2, 1.5]$. Note that the first item does not appear in the rule since they have the value 01 in the two tag bits

Focusing on the genetics operators, the algorithm select a set of individuals to act as parents (see line 14, Fig. 6.8) by using the roulette-wheel selection [12]. This method assumes that the probability of selection of an individual is proportional to its fitness value when all the fitness values of the remaining individuals are considered. Let us consider n individuals where the fitness value of a specific individual i is denoted as f_i. The selection probability of an individual i is given by the probability $p_i = f_i / \sum_{j=1}^{n} f_j$. Once a set of individuals have been chosen by the selection method, new individuals are generated thanks to some genetic operators (see line 15, Fig. 6.8). The crossover operator used in this algorithm is multiple point crossover, which randomly chooses n cut points from the chromosomes of two parents and the obtained substrings are then swapped. As for the mutation operator, it is a simple operator that randomly modifies the two tag bits of an item.

Another interesting multiobjective approach was proposed by Minaei-Bidgoli et al. [24], which considered the same model as the one proposed by Ghosh et al. [11] but introducing the concept of continuous patterns. Thus, the algorithm encodes rules as linear chromosomes, and each gene comprises two values (lower and upper bound of a continuous attribute) and two extra tag bits. For example, Fig. 6.9 shows a sample individual that represents the rule $Att_3[3.3, 3.6] \wedge Att_4[0.1, 0.6] \rightarrow Att_2[1.2, 1.5]$. We assumed that only four attributes exist in the dataset.

Other authors have also considered the extraction of continuous patterns by means of a new multiobjective evolutionary model, which searches for solutions having a good tradeoff between three objectives: (1) comprehensibility defined as the inverse of the number of attributes within the rule; (2) lift; and (3) the performance metric [22]. In this work, performance is defined as the product of support and certainty factor, which was defined and analysed in depth in Chap. 2. This multiobjective evolutionary algorithm, known as QAR-CIP-NSGA-II, extends the well-known MOEA Non-dominated Sorting Genetic Algorithm II (NSGA-II) [8] to perform an evolutionary learning of the intervals of the continuous attributes.

QAR-CIP-NSGA-II considers each chromosome as a string of genes where each gene defines the lower and upper bounds of a continuous attribute, and the i-th attribute is encoded in the i-th gene used (see Fig. 6.10). Additionally, each gene includes an additional part to represent whether the specific gene is included or not in the rule. When this part is equal to -1, it means that the attribute is not included in the rule, whereas the values 0 and 1 describe that the attribute is included in the antecedent or consequent of the rule, respectively. Once the individuals are encoded, the algorithm follows an iterative process as shown in Fig. 6.11.

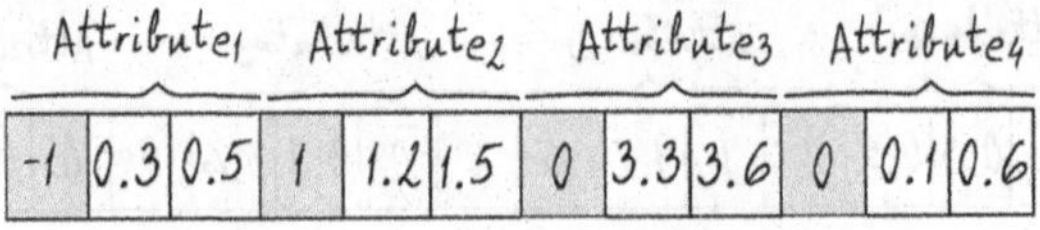

$$\text{Attribute}_1 \quad \text{Attribute}_2 \quad \text{Attribute}_3 \quad \text{Attribute}_4$$

-1	0.3	0.5	1	1.2	1.5	0	3.3	3.6	0	0.1	0.6

Fig. 6.10 Example of an individual that represents the rule $Att_3[3.3, 3.6] \wedge Att_4[0.1, 0.6] \rightarrow Att_2[1.2, 1.5]$. Note that the first item does not appear in the rule since they have the value -1 in the first part of the gene

Require: *population_size* {Number of individuals and generations considered}
Ensure: *population*
1: *population* ← ∅
2: *parents* ← ∅
3: *offspring* ← ∅
4: *pareto_front* ← ∅
5: *population* ← generate(*population_size*)
6: **for** ∀*member* ∈ *population* **do**
7:      evaluate(*member*)
8: **end for**
9: **while** termination condition is not reached **do**
10:     generate all non-dominated fronts $F = (F_1, F_2, ...)$ from population and calculate crowding-
        distance in F_i
11:     *pareto_front* ← F_1
12:     $i \leftarrow 2$
13:     **while** $|population| + |F_i| \leq population\_size$ **do**
14:         *population* ← *population* ∪ F_i
15:         *i*++
16:     **end while**
17:     *parents* ← obtain a set of individuals from *population* by a tournament selector
18:     *offspring* ← apply genetic operators on the set *parents*
19:     **for** ∀*member* ∈ *population* **do**
20:         evaluate(*member*)
21:         *population* ← update(*population*,*offspring*)
22:     **end for**
23: **end while**

Fig. 6.11 Pseudo-code of the QAR-CIP-NSGA-II algorithm [22]

In this algorithm, a set of individuals is randomly selected to act as parents by using a tournament selector (see line 18, Fig. 6.11. Then, a set of genetic operators are applied to the parents (see line 19, Fig. 6.11). The crossover operator generates two offspring by randomly interchanging the genes of the parents. The mutation operator consists of randomly modifying the lower and upper intervals of a gene selected at random. Additionally, the mutation operator is able to change the part of the gene that determines whether the attribute is included in the antecedent, in the consequent, or not considered for the rule.

Finally, it should be noted that most of the MOEAs for mining pattern focus on positive dependencies without paying particular attention to negative dependencies, which relate the presence of certain items to the absence of others. To this

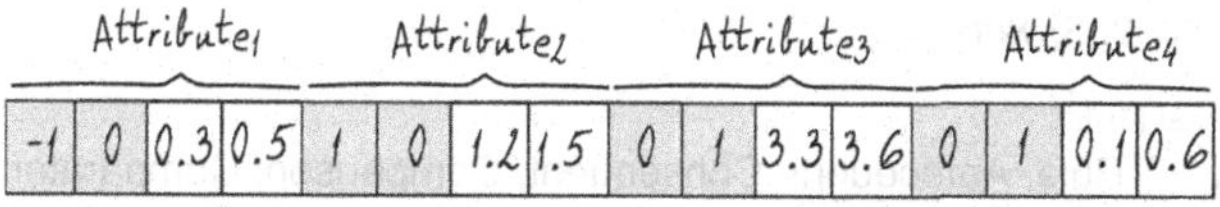

Attribute₁				Attribute₂				Attribute₃				Attribute₄			

-1	0	0.3	0.5	1	0	1.2	1.5	0	1	3.3	3.6	0	1	0.1	0.6

Fig. 6.12 Example of an individual that represents the rule $Att_3[3.3, 3.6] \wedge Att_4[0.1, 0.6] \rightarrow \neg Att_2[1.2, 1.5]$. Note that the first item does not appear in the rule since they have the value -1 in the first part of the gene

end, Martín et al. [21] proposed the MOPNAR algorithm, a new multiobjective evolutionary algorithm that mines a reduced set of positive and negative quantitative association rules with low computational cost.

MOPNAR considers each chromosome as a string of genes where each gene defines the lower and upper bounds of a continuous attribute. Additionally, each gene comprises two extra values: (1) a value that determines whether the attribute will be included or not in the rule, and in which part of the rule will be located (antecedent or consequent); and (2) a value that indicates whether the specific attribute is positive or negative. Thus, if the first value is equal to -1, it means that the attribute is not included in the rule, whereas the values 0 and 1 describe that the attribute is included in the antecedent or consequent of the rule, respectively. If the second value is equal to 1, then it means the attribute appears as a positive dependency, whereas a 0 value indicates a negative dependency. Figure 6.12 illustrates a sample individual encoded by MOPNAR.

Analysing the objectives to be optimized by MOPNAR, it is obtained that this algorithm uses the same objectives as QAR-CIP-NSGA-II [22]. Thus, the aim is to obtain a set of solutions that achieve a good tradeoff between: (1) comprehensibility defined as the inverse of the number of attributes within the rule; (2) lift; and (3) the performance metric, which is defined as the product of support and certainty factor.

Finally, it should be noted that MOPNAR was described as an extension of a MOEA based on decomposition MOEA/D-DE [16], which decomposes the multiobjective optimization problem into N scalar optimization subproblems and uses an evolutionary algorithm to optimize these subproblems simultaneously. In order to store all the nondominated rules found, provoke diversity in the population, and improve the coverage of the dataset, MOPNAR makes use of both an external population and a restarting process. The external population enables all the nondominated rules found to be stored, including a procedure to avoid overlapping rules.

6.3.2 Genetic Programming

The use of grammars for mining patterns of interest was first described by Luna et al. [18]. This algorithm, known as G3PARM (Grammar-Guided Genetic Programming for Association Rule Mining), defines a grammar where solutions comprise a

G = (Σ_N, Σ_T, P, S) with:

> S   = Rule
> Σ_N = {Rule, Antecedent, Consequent, Comparison, Comparator_Discrete,
>   Comparator_Continuous, Attribute_Discrete, Attribute_Continuous}
> Σ_T = {'∧', '=', '≠', '<', '≤', '>', '≥', 'name', 'value'}
> P   = {Rule = Antecedent, Consequent ;
>   Antecedent = Comparison | '∧', Comparison, Antecedent ;
>   Consequent = Comparison ;
>   Comparison = Discrete, Attribute | Continuous, Attribute ;
>   Discrete = '≠' | '=' ;
>   Continuous = '<' | '≤' | '>' | '≥' ;
>   Attribute = 'name', 'value' ;}

Fig. 6.13 Context-free grammar defined for the G3PARM algorithm considering positive, negative, discrete and continuous patterns

Table 6.1 Sample metadata

Attributes	Values
Toy	Ball, teddy, doll
Toy's price	[100, 1500]
Sex	Male, female

set of items in the antecedent of the rule, whereas the consequent is formed only by a single item. The use of grammars enables both positive and negative patterns to be obtained, and only a simple modification in the grammar provokes the extraction of either positive or negative patterns.

Due to the promising results and features of G3PARM, some authors have proposed different extensions of this algorithm founded on two well-known multiobjective algorithms: NSGA-II [8] and SPEA2 [39]. One of these algorithms is known as NSGA-G3PARM [20] and they share some characteristics from both G3PARM and NSGA-II. For example, the encoding criterion is quite similar to the one proposed in G3PARM [18] with the only difference that this new encoding criterion is able to represent continuous patterns. Thus, each individual is encoded as a sentence of the language generated by the grammar G (see Fig. 6.13), considering a maximum number of derivations to avoid extremely large trees. Thus, considering the grammar defined in this approach, the following language is obtained $L(G) =$ { Comparison ($\wedge$ Comparison)^n $\rightarrow$ Comparison , $n \geq 0$}. To obtain individuals, a set of derivation steps is carried out by applying the production rules declared in the set P and starting from the start symbol *Rule*. It should be noted that each terminal symbol randomly chooses the names (*name*) and values (*value*) from the dataset metadata, so considering the sample metadata illustrated in Table 6.1, the terminal symbol *name* adopts any value from the following: *toy*, *toy's price*, or *sex*. Once the symbol *name* is assigned, a random value is selected in its domain, e.g., *ball*, *teddy*, or *doll* for the attribute *toy*. Thus, each individual is represented by a syntax-tree structure according to the defined grammar, and considering a maximum depth to avoid infinite derivations. Figure 6.14 illustrates a sample individual encoded

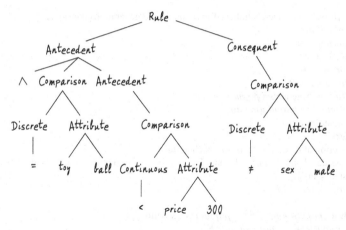

Fig. 6.14 Sample individual encoded by using the grammar shown in Figure 6.13

from the grammar G (see Fig. 6.13) and the sample metadata (see Table 6.1). The individual represented by the derivation syntax-tree represents the rule **IF** *toy = ball* ∧ *price = cheap* **THEN** *sex ≠ male*.

Focusing on the pseudocode of the NSGA-G3PARM algorithm, it is shown in Fig. 6.15. In this proposal, different measures, which serve to determine the quality of the rules mined, are used as objective functions. Authors proposed a study of two pairs of measures, so either support-confidence and support-lift are optimized, and a posteriori analysis determines which pair of measures obtains the best pareto front based on SP and HV quality measures (see Sect. 6.2.2). In each iteration of the evolutionary process (lines 10–24, Fig. 6.15), NSGA-G3PARM selects a set of individuals to act as parents by means of a tournament selector, and this set of parents are crossed and mutated in order to obtain new individuals (line 19, Fig. 6.15). The crossover and mutation genetic operators are the same used by G3PARM and described in [19], which are based on the fact that highly frequent items might produce frequent patterns and, therefore, frequent association rules. The main idea of these genetic operators is to replace the least frequent item defined within a parent to produce a new more frequent item. Thus, the crossover operator removes the least frequent item from a parent and it is replaced by the most frequent item from another parent. The same concept is considered for the mutation operator, which replaces the least frequent item from a parent to produce a completely new random subtree. The new set of individuals generated by the genetic operators are then evaluated to determine the values of the quality measures (lines 8 and 21). Since the ultimate goal is to return a pre-defined number of optimal solutions, individuals have to be organized in fronts (line 11). Therefore, the algorithm identifies those solutions that are not dominated by any other. After that, the process is repeated on the remaining solutions, so new fronts are calculated (see line 11). Additionally, it is useful to determine the density of the solutions surrounding a particular solution, so it is possible to obtain the best solution in each front. To this end, the average

Require: *population_size* {Number of individuals and generations considered}
Ensure: *population*
 1: *population* ← ∅
 2: *parents* ← ∅
 3: *offspring* ← ∅
 4: *pareto_front* ← ∅
 5: *population* ← generate(*population_size*)
 6: **for** ∀*member* ∈ *population* **do**
 7:     evaluate(*member*)
 8: **end for**
 9: **while** termination condition is not reached **do**
10:     generate all non-dominated fronts $F = (F_1, F_2, ...)$ from population and calculate crowding-distance in F_i
11:     *pareto_front* ← F_1
12:     $i ← 2$
13:     **while** $|population| + |F_i| \leq population\_size$ **do**
14:         *population* ← *population* ∪ F_i
15:         i++
16:     **end while**
17:     *parents* ← obtain a set of individuals from *population* by a tournament selector
18:     *offspring* ← apply genetic operators on the set *parents*
19:     **for** ∀*member* ∈ *population* **do**
20:         evaluate(*member*)
21:         *population* ← update(*population*, *offspring*)
22:     **end for**
23: **end while**

Fig. 6.15 Pseudo-code of the NSGA-G3PARM algorithm [20]

distance to each solution around each of its objectives is calculated. Those solutions having the highest and lowest values of each objective are assigned an infinite distance value. On the other hand, intermediate solutions are assigned a distance value equal to the absolute normalized difference in the objective function values of two adjacent solutions. Finally, the overall distance value for each solution is calculated as the sum of the distance values for each objective function.

Another interesting algorithm based on grammars is SPEA-G3PARM [20], which is an adaptation of G3PARM [18] to the well-known SPEA-2 [39] multi-objective algorithm. This algorithm presents some similar features to the NSGA-G3PARM algorithm previously described. For example, the encoding criterion is the same, so individuals are encoded by using the grammar defined in Fig. 6.13. Additionally, the same genetic operators are considered. The pseudocode of this algorithm is shown in Fig. 6.16. The algorithm starts by obtaining the set *population* of individuals (see line 7, Fig. 6.16) by using the grammar illustrated in Fig. 6.13. The SPEA-G3PARM algorithm follows an evolutionary strategy, so the set *population* evolves through the generations (lines 11–26), generating new individuals by means of genetic operators (line 21). The main characteristic of this algorithm is that each individual is evaluated according to the number of population members that are dominated by or equal to the specific individual with respect to the objective values,

Require: *population_size, pareto_size* {Number of individuals and generations considered}
Ensure: *pareto_front*
 1: *population* ← ∅
 2: *parents* ← ∅
 3: *offspring* ← ∅
 4: *pareto_front* ← ∅
 5: *population* ← generate(*population_size*)
 6: **for** ∀*member* ∈ *population* **do**
 7:     evaluate *member* according to a function based on dominance
 8: **end for**
 9: **while** termination condition is not reached **do**
10:     *pareto_front* ← obtain the pareto optimal front from the set *population* ∪ *pareto_front*
11:     **if** size(*pareto_front*) > *pareto_size* **then**
12:         *pareto_front* ← reduce the set *pareto_front* to the size of *pareto_size*
13:     **else**
14:         **if** size(*pareto_front*) < *pareto_size* **then**
15:             *pareto_front* ← add new solutions to *pareto_front* from the set *population*
16:         **end if**
17:     **end if**
18:     *parents* ← obtain a set of individuals from *population* by a tournament selector
19:     *offspring* ← applyGeneticOperators(*parents*)
20:     **for** ∀*member* ∈ *population* **do**
21:         evaluate *member* according to a function based on dominance
22:     **end for**
23:     *population* ← update(*population, offspring*)
24: **end while**

Fig. 6.16 Pseudo-code of the SPEA-G3PARM algorithm [20]

divided by the population size plus one. In each generation, the algorithm obtains the pareto optimal front (line 12). If the size of this front is greater than a pre-defined size, then the front has to be downsized by choosing the best individuals ranked according to the fitness function (lines 13–15). On the contrary, if the size of the front is lower than the pre-defined size, then it is necessary to fill it with the best solutions from the second front (lines 16–18).

6.3.3 Other Algorithms

Alatas et al. [3] proposed a pareto-based multiobjective differential evolution algorithm as a search strategy for mining accurate and comprehensible continuous patterns and associations between the patterns. In this algorithm, known MOD-ENAR, individuals consist of decision variables which represent the items and intervals—the *i*-th item is encoded in the *i*-th decision variable. Each decision variable has three parts. The first part is used to represent whether the variable describes the antecedent, the consequent, or is not included in the rule. A 0 value is used to indicate that this item will be in the antecedent of the rule; the value 1

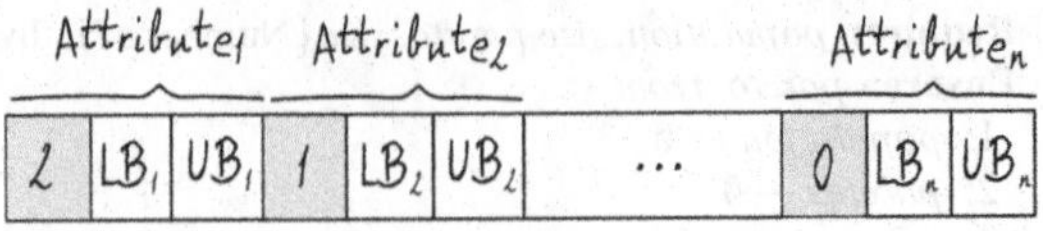

Fig. 6.17 Sample individual representation for the differential evolution algorithm proposed in [3]

is used to indicate the consequent; whereas a 2 value means that the item will not be involved in the rule. Finally the second and third parts of any decision variable represent the lower and upper bounds of the item interval, respectively. The structure of an individual has been illustrated in Fig. 6.17, where n is the number of attributes of data being mined.

The MODENAR algorithm considers four quality measures to be optimized. As traditional algorithms for mining association rules, MODENAR considers both support and confidence as good quality measures to determine the interest of the rules mined. Additionally, this algorithm uses comprehensibility, which is related to the number of attributes involved in both antecedent and consequent part of the rule. According to the authors, the motivation behind this objective is to bias the algorithm towards slightly shorter rules, so readability is also increased. Finally, MODENAR also considers the amplitude of the intervals as a metric to be optimized. Thus, if two individuals cover the same set of transactions, the one whose intervals are smaller gives us better information. Note that support, confidence and comprehensibility are maximization objectives, whereas the amplitude is a minimization objective.

Finally, another interesting multiobjective optimization algorithm for mining patterns of interest (MOGBAP-ARM) was proposed by Olmo et al. [26], which is based on an ant programming technique that has recently reported very promising results in other rule based tasks. This algorithm considers support and confidence as quality measures to calculate the importance of each solution. Once these two measures are calculated for all the individuals, a fitness value is obtained for each one using ranks. Thus, population is organized in fronts of non-dominance similarly to the procedure used by the NSGA2 algorithm [8]. The fitness assigned to a given individual corresponds to the inverse value of the front to which it belongs, so individuals from the first front will receive a fitness function value of 1, whereas individuals from the second front will be assigned with a value of 0.5. Additionally, as in any ant programming algorithm, MOGBAP-ARM follows a reinforcement and evaporation process. In this algorithm, the amount of pheromones that a given ant will deposit in its path is inversely proportional to the front number to which this individual belongs.

The MOGBAP-ARM algorithm works as follows. It starts by creating an empty space of states, and a set of ants, and then, it loops through the following processes until it has performed maximum number of iterations. Each ant is evaluated by the objective functions (support and confidence), and the states visited by each ant are kept in the data structure that represents the space of states. All individuals are ranked according to the pareto dominance, so that individuals belonging to front N

will receive a fitness value equal to $1/N$. Then, all individuals reinforce the amount of pheromone in their path in proportion to their fitness values, and each transition in the path of a given ant perceiving an equal amount of pheromones.

6.4 Successful Applications

Nowadays, there is a considerable increase in the number of applications of MOEAs, and it is mainly originated by the success of MOEAs in solving real-world problems, obtaining better results than those produced using other search techniques. Engineering is probably the most popular application domain area within the MOEAs literature [29]. This is mainly because engineering applications normally define mathematical models that can directly be associated with a MOEA search. Scientific applications [33] are also highly studied by the research community in the use of MOEAs.

Since the pattern mining task requires more than a single quality measure to be optimized, the use of multiobjective optimization in pattern mining has also been considered by different researchers in different application domains. One of these application domains is molecular biology, where one of the main goals is the reconstruction of gene regulatory processes by the inference of regulatory interactions among genes. In this regard, the use of pattern mining helps to infer the relationships between genes from an organism in a particular biological process [23]. To do so, authors proposed the GarNet (Gene-gene associations from Association Rules for inferring gene NETworks) algorithm, which is based on the well-known multiobjective evolutionary approach NSGA-II [8] to discover continuous patterns without performing a previous discretization.

References

1. A. Abraham, L.C. Jain, and R. Goldberg. *Evolutionary Multiobjective Optimization: Theoretical Advances and Applications*. Advanced Information and Knowledge Processing. Springer, 2005.
2. C. C. Aggarwal and J. Han. *Frequent Pattern Mining*. Springer International Publishing, 2014.
3. B. Alatas, E. Akin, and A. Karci.   MODENAR: Multi-objective Differential Evolution Algorithm for Mining Numeric Association Rules. *Applied Soft Computing*, 8:646–656, 2008.
4. S. Brin, R. Motwani, J. D. Ullman, and S. Tsur. Dynamic Itemset Counting and Implication Rules for Market Basket Data. In *Proceedings of the 1997 ACM SIGMOD International Conference on Management of Data*, SIGMOD '97, pages 255–264, Tucson, Arizona, USA, 1997. ACM.
5. C. A. Coello, G. B. Lamont, and D. A. Van Veldhuizen.   *Evolutionary Algorithms for Solving Multi-Objective Problems (Genetic and Evolutionary Computation)*. Springer-Verlag New York, Inc., Secaucus, NJ, USA, 2006.
6. L. Davis. *Handbook of Genetic Algorithms*. Van Nostrand Reinhold, 1991.

7. K. Deb. *Multi-Objective Optimization Using Evolutionary Algorithms*. John Wiley & Sons, Inc., New York, NY, USA, 2001.
8. K. Deb, A. Pratap, S. Agrawal, and T. Meyarivan. A Fast Elitist Multi-Objective Genetic Algorithm: NSGA-II. *IEEE Transactions on Evolutionary Computation*, 6:182–197, 2000.
9. Carlos M. Fonseca and Peter J. Fleming. Genetic algorithms for multiobjective optimization: Formulation discussion and generalization. In *Proceedings of the 5th International Conference on Genetic Algorithms*, pages 416–423, San Francisco, CA, USA, 1993. Morgan Kaufmann Publishers Inc.
10. L. Geng and H. J. Hamilton. Interestingness Measures for Data Mining: A Survey. *ACM Computing Surveys*, 38, 2006.
11. A. Ghosh and B. Nath. Multi-objective Rule Mining Using Genetic Algorithms. *Information Science*, 163(1–3):123–133, 2004.
12. D. E. Goldberg. *Genetic Algorithms in Search, Optimization and Machine Learning*. Addison-Wesley Longman Publishing Co., Inc., Boston, MA, USA, 1st edition, 1989.
13. F. Guillet and H. Hamilton. *Quality Measures in Data Mining*. Springer Berlin / Heidelberg, 2007.
14. H. Ishibuchi, I. Kuwajima, and Y. Nojima. Multiobjective association rule mining. In *Proceedings of the Multiobjective Problem Solving from Nature*, Reykjavik, Iceland, 2006.
15. A. Jiménez, F. Berzal, and J. C. Cubero. Interestingness measures for association rules within groups. In *Proceedings of the 13th International Conference on Information Processing and Management of Uncertainty in Knowledge-Based Systems*, IPMU 2010, pages 298–307. Springer, 2010.
16. H. Li and Q. Zhang. Multiobjective optimization problems with complicated pareto sets, moea/d and nsga-ii. *IEEE Transactions on Evolutionary Computation*, 13(2):284–302, 2009.
17. B. Liu, W. Hsu, S. Chen, and Y. Ma. Analyzing the subjective interestingness of association rules. *IEEE Intelligent Systems*, 15(5):47–55, 2000.
18. J. M. Luna, J. R. Romero, and S. Ventura. G3PARM: A Grammar Guided Genetic Programming Algorithm for Mining Association Rules. In *Proceedings of the IEEE Congress on Evolutionary Computation*, IEEE CEC 2010, pages 2586–2593, Barcelona, Spain, 2010.
19. J. M. Luna, J. R. Romero, and S. Ventura. Design and behavior study of a grammar-guided genetic programming algorithm for mining association rules. *Knowledge and Information Systems*, 32(1):53–76, 2012.
20. J. M. Luna, J. R. Romero, and S. Ventura. Grammar-based multi-objective algorithms for mining association rules. *Data & Knowledge Engineering*, 86:19–37, 2013.
21. D. Martín, A. Rosete, J. Alcalá, and F. Herrera. A new multiobjective evolutionary algorithm for mining a reduced set of interesting positive and negative quantitative association rules. *IEEE Transactions on Evolutionary Computation*, 18(1):54–69, 2014.
22. D. Martín, A. Rosete, J. Alcalá-Fdez, and F. Herrera. Qar-cip-nsga-ii: A new multi-objective evolutionary algorithm to mine quantitative association rules. *Information Sciences*, 258:1–28, 2014.
23. M. Martínez-Ballesteros, I. A. Nepomuceno-Chamorro, and J. C. Riquelme. Discovering gene association networks by multi-objective evolutionary quantitative association rules. *Journal of Computer System Sciences*, 80(1):118–136, 2014.
24. B. Minaei-Bidgoli, R. Barmaki, and M. Nasiri. Mining numerical association rules via multi-objective genetic algorithms. *Information Sciences*, 233:15–24, 2013.
25. J. L. Olmo, J. M. Luna, J. R. Romero, and S. Ventura. Association rule mining using a multi-objective grammar-based ant programming algorithm. In *Proceedings of the 11th International Conference on Intelligent Systems Design and Applications*, ISDA 2011, pages 971–977, Cordoba, Spain, 2011.
26. J. L. Olmo, J. M. Luna, J. R. Romero, and S. Ventura. Mining association rules with single and multi-objective grammar guided ant programming. *Integrated Computer-Aided Engineering*, 20(3):217–234, 2013.

27. G. Piatetsky-Shapiro. Discovery, analysis and presentation of strong rules. In G. Piatetsky-Shapiro and W. Frawley, editors, *Knowledge Discovery in Databases*, pages 229–248. AAAI Press, 1991.
28. H. R. Qodmanan, M. Nasiri, and B. Minaei-Bidgoli. Multi objective association rule mining with genetic algorithm without specifying minimum support and minimum confidence. *Expert Systems with Applications*, 38:288–298, 2011.
29. P. M. Reed, B. S. Minsker, and D. E. Goldberg. Designing a new elitist nondominated sorted genetic algorithm for a multiobjective long term groundwater monitoring application. In *Proceedings of the 2001 Genetic and Evolutionary Computation Conference*, GECCO 2001, pages 352–356, San Francisco, California, 2001.
30. T. Scheffer. Finding association rules that trade support optimally against confidence. In *Proceedings of the 5th European Conference of Principles and Practice of Knowledge Discovery in Databases*, PKDD 2001, pages 424–435, Freiburg, Germany, 2001.
31. P. D. Shenoy, K. G. Srinivasa, K. R. Venugopal, and L. M. Patnaik. Dynamic association rule mining using genetic algorithms. *Intelligent Data Analysis*, 9(5):439–453, 2005.
32. W. Stadler. A survey of multicriteria optimization or the vector maximum problem, part i: 1776–1960. *Journal of Optimization Theory and Applications*, 29(1):1–52, 1979.
33. T. J. Stewart, R. Janssen, and M. van Herwijnen. A genetic algorithm approach to multiobjective land use planning. *Computers and Operations Research*, 31(14):2293–2313, 2004.
34. P. Tan and V. Kumar. Interestingness Measures for Association Patterns: A Perspective. In *Proceedings of the Workshop on Postprocessing in Machine Learning and Data Mining*, KDD '00, New York, USA, 2000.
35. D. A. Van Veldhuizen. *Multiobjective Evolutionary Algorithms: Classifications, Analyses, and New Innovations*. PhD thesis, Wright Patterson AFB, OH, USA, 1999. AAI9928483.
36. D. A. Van Veldhuizen and G. B. Lamont. Multiobjective evolutionary algorithm research: A history and analysis. Technical report, Technical Report TR-98-03, Department of Electrical and Computer Engineering, Graduate School of Engineering, Air Force Institute of Technology, Wright-Patterson AFB, Ohiod, 1998.
37. C. Zhang and S. Zhang. *Association rule mining: models and algorithms*. Springer Berlin / Heidelberg, 2002.
38. E. Zitzler, K. Deb, and L. Thiele. Comparison of multiobjective evolutionary algorithms: Empirical results. *Evolutionary Computation*, 8(2):173–195, 2000.
39. E. Zitzler, M. Laumanns, and L. Thiele. SPEA2: Improving the Strength Pareto Evolutionary Algorithm for Multiobjective Optimization. In *Proceedings of the 2001 conference on Evolutionary Methods for Design, Optimisation and Control with Application to Industrial Problems*, EUROGEN 2001, pages 95–100, Athens, Greece, 2001.

Chapter 7
Supervised Local Pattern Mining

Abstract Pattern mining is considered as a really interesting task for the extraction of hidden knowledge in the form of patterns. The extraction of such subsequences, substructures or itemsets that represent any type of homogeneity and regularity in data has been carried out from unlabeled data. However, there are many research areas that aim at discovering patterns in the form of rules induced from labeled data. Hence, it is interesting to discover patterns and associations from a supervised point of view since a single item or a set of them can be considered as distinctive. This task, which is known as supervised local pattern mining, is described in this chapter, including different areas such as contrast set mining, emerging pattern mining, and subgroup discovery. This chapter is mainly focused on the subgroup discovery task, which is widely known in the field of supervised local pattern mining. Here, an exhaustive description about this task is provided, including some important quality measures in the field. Then, this chapter includes different evolutionary approaches for subgroup discovery. Finally, this chapter includes an analysis of other different approaches proposed for mining and quantifying local patterns.

7.1 Introduction

Traditional pattern mining models are based on the extraction of hidden knowledge from unlabeled data [27]. These models encompass techniques that seek for intrinsic and important properties of datasets, i.e. subsequences, substructures or itemsets that represent any type of homogeneity and regularity in data [2]. These models can also be applied to a supervised context [50], where a single item or a set of them is considered as interesting and distinctive, and their goal is to obtain relationships between these items and the others. Pattern mining methods applied to supervised contexts are grouped under the heading of *supervised local pattern mining*, which aims to mine relationships that can be much more appropriate in many different fields [5, 51]. It should be noted that the description and analysis obtained by these methods are more targeted and useful, so there is a growing interest in this kind of patterns and associations giving rise to different areas such as contrast set mining [8], emerging pattern mining [19], subgroup discovery [13], etc.

Patterns considered by these models are represented in the form of rules where the conditions included in the antecedent are used to identify transactions for which a certain property in the target item (or set of items) holds. Thus, pattern discovery in the supervised local pattern mining field is a task that lies at the intersection of predictive and descriptive induction [43]. Despite the fact that the discovery of such patterns can be considered as a form of supervised and predictive learning, it is also a form of extracting descriptive knowledge since the goal is to discover interesting associations between the patterns. Thus, the mining of local patterns aims at describing groups by means of independent and simple rules unlike predictive learning which aims at forming complex models that explain future behaviour.

Supervised local pattern mining is the task of finding patterns that describe subsets of data with a high statistical unusualness in their distribution, and these subsets are formed by using a single target feature or a set of them. Different approaches for supervised local pattern mining have been proposed, which mainly differ in the way in which the statistical unusualness is quantified. One of the most widely-known supervised local pattern mining techniques is subgroup discovery [29], which attempts to search for relations in the form of rules between different properties or variables of a set with respect to a target variable. The main difference against classification tasks is that subgroup discovery seeks to extract interesting subgroups (not an accurate classifier) for a given target feature and the subgroups discovered do not cover all the transactions for the target feature (see Fig. 7.1).

Early subgroup discovery approaches were based on deterministic models, consisting in adapting either association rule mining or classification algorithms for the subgroup discovery task [32]. Then, many authors [12, 17] have proposed different evolutionary algorithms to reduce the computational time and to increase the interpretability of the resulting model. Additionally, some other authors [39] have also proposed the use of grammars to include external and subjective knowledge to the mining process. A major feature of any of the evolutionary algorithms proposed to this aim is their ability to discover reliable and highly representative subgroups with a low number of rules and variables per rule.

Fig. 7.1 Difference between classification and subgroup discovery tasks

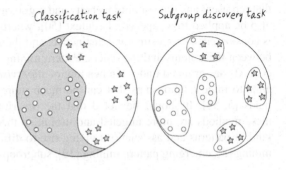

7.2 Subgroup Discovery

Subgroup discovery [6] has been established as a really versatile and effective technique in data mining thanks to its descriptive and exploratory nature. The aim of this interesting technique for mining supervised local patterns is to identify interesting groups of patterns according to their distributional unusualness with respect to a certain property of interest. In this section, the formal definition of subgroup discovery is firstly described. Then, different quality measures are formally defined. Finally, this section also deals with different algorithms in this field.

7.2.1 Problem Definition

Subgroup discovery is considered as a broadly applicable data mining technique whose aim is to discover some interesting relationships between patterns or variables in a set with respect to a specific target variable. The information extracted by this data mining technique is normally represented in the form of rules according to the user's interest.

The concept of subgroup discovery was first introduced by Klösgen [34] and Wrobel [48] as follows: *"Given a population of individuals (customers, objects, etc.) and a property of those individuals that we are interested in, the task of subgroup discovery is to find population subgroups that are statistically most interesting for the user, e.g., subgroups that are as large as possible and have the most unusual statistical characteristics with respect to a target attribute of interest"*. Analysing the subgroup discovery technique, it is obtained that it combines the features of supervised and unsupervised learning tasks [10] and uncovers explicit subgroups via single and simple rules. Thus, it is important for any subgroup discovery algorithm to mine interesting subgroups by means of rules having a clear structure and few variables [29]. Those rules are also required to be of high interest and cover as many instances of the class attribute as they can do.

In subgroup discovery, a rule R, which consists of an induced subgroup description, can be formally represented as a relation of the form $X \rightarrow Target$. The antecedent X of the rule is a conjunction of features that describes an unusual statistical distribution with respect to the consequent, which is identified by a target value for a variable of interest. Subgroup discovery has been well investigated concerning binary and nominal target variables, i.e., properties of interest with a finite number of possible values [6]. Nevertheless, this is not the only type of target variables that can be used in subgroup discovery so the use of target variables defined in continuous domains has received increasing attention recently [4, 26, 41]. Additionally, subgroup discovery not only might include binary, discrete, and continuous target variables, but it also extends to multi-target concepts [21], and to exceptional model mining [36].

7.2.2 Quality Measures

The choice of good quality measures in subgroup discovery has been discussed by many researchers [20], resulting in a wide variety of metrics. However, there is no clear consensus about which specific quality measure is the right to each particular problem. Some studies about this problem have given rise to a classification of the measures [29], describing four main groups: measures of complexity, generality, precision and interest. Analysing the first group, which includes complexity measures, it is obtained that these metrics are highly related to the interpretability of the subgroups discovered. Thus, this group of measures quantifies the simplicity of the knowledge extracted from the discovered subgroups. Two measures in this regard are the number of rules (defining subgroups) and the average number of variables in the antecedent, which is computed as the average number of variables for each rule within the set of induced rules.

As for the generality measures, this group of measures quantifies the quality of the subgroups according to the patterns covered. One of the most commonly used measures in this sense is support, which is a well-known quality measure used in association rule mining. This measure calculates the number of transactions covered by the rule on the basis of the whole set of transactions included in the database (see Eq. (7.1)). A modified version of the support quality measure is the support on the basis of the transactions of the target feature. This measure is formally described in Eq. (7.2) as the fraction of retrieved transactions T that are relevant, i.e., the percentage of transactions from the dataset $\mathcal{D}$ that satisfy the antecedent X and the consequent (the target feature) on the basis of those satisfied by the consequent.

$$support(X \rightarrow Target) = \frac{|\{X \cup Target \subseteq T, T \in \mathcal{D}\}|}{|\mathcal{D}|} \tag{7.1}$$

$$support_{Target}(X \rightarrow Target) = \frac{|\{X \cup Target \subseteq T, T \in \mathcal{D}\}|}{|\{Target \subseteq T, T \in \mathcal{D}\}|} \tag{7.2}$$

Another quality metric defined as a generality measure is the coverage (see Eq. (7.3)), which calculates the percentage of transactions of the dataset covered by the antecedent part of the rule over the whole set of transactions in the dataset.

$$support(X \rightarrow Target) = \frac{|\{X \subseteq T, T \in \mathcal{D}\}|}{|\mathcal{D}|} \tag{7.3}$$

Focusing on the group of measures related to the precision property, one of the most commonly used measures is the confidence of the rule. This measure determines the reliability of a subgroup, measuring the relative frequency of examples that satisfy the complete rule among those satisfying only the antecedent X (see Eq. (7.4)). It is interesting to note that this quality measure can also be found as accuracy in the specialised bibliography.

$$confidence(X \rightarrow Target) = \frac{|\{X \cup Target \subseteq T, T \in \mathscr{D}\}|}{|\{X \subseteq T, T \in \mathscr{D}\}|} \tag{7.4}$$

Precision measure Q_c is a metric that quantifies the tradeoff between the true and false positives covered in a lineal function, and it can be computed as shown Eq. (7.5). Note that $|\{X \cup \overline{Target} \subseteq T, T \in \mathscr{D}\}|$ is the number of transactions satisfying the antecedent of the rule but not its target variable, whereas the parameter c is used as a generalisation constant.

$$Q_c(X \rightarrow Target) = |\{X \cup Target \subseteq T, T \in \mathscr{D}\}| - c \times |\{X \cup \overline{Target} \subseteq T, T \in \mathscr{D}\}| \tag{7.5}$$

Precision measure Q_g is another precision quality metric that measures the tradeoff of a subgroup between the number of transactions classified perfectly and the unusualness of their distribution (see Eq. (7.6)). Note that g is used as a generalisation parameter that usually takes values between 0.5 and 100.

$$Q_g(X \rightarrow Target) = \frac{|\{X \cup Target \subseteq T, T \in \mathscr{D}\}|}{|\{X \cup \overline{Target} \subseteq T, T \in \mathscr{D}\}| + g} \tag{7.6}$$

Finally, the group of measures of interest comprises measures such as significance, novelty, interest, etc. Significance measure indicates the significance of a finding measured by the likelihood ratio of a rule, where n_T is computed as the number of targets in Eq. (7.7). As for the novelty, this is a metric defined to detect unusual subgroups (see Eq. (7.8))

$$significance(X \rightarrow Target) = 2 \times \sum_{i=1}^{n_T} (|\{X \cup Target_i \subseteq T, T \in \mathscr{D}\}|) \tag{7.7}$$

$$\times \log \frac{|\{X \cup Target_i \subseteq T, T \in \mathscr{D}\}|}{|\{X \subseteq T, T \in \mathscr{D}\}| \times \frac{|\{Target_i \subseteq T, T \in \mathscr{D}\}|}{|\mathscr{D}|}}$$

$$Novelty(X \rightarrow Target) = |\{X \cup Target \subseteq T, T \in \mathscr{D}\}| \tag{7.8}$$
$$-(|\{X \subseteq T, T \in \mathscr{D}\}| \times |\{Target \subseteq T, T \in \mathscr{D}\}|)$$

Despite the fact that any quality measure in the subgroup discovery field should be located in one of the aforementioned groups of measure, it is also possible to fined some quality measures that lies somewhere between generality, interest and precision. One example of that kind of measures is the unusualness of a rule, defined as the weighted relative accuracy and computed as depicted in Eq. (7.9).

$$unusualness(X \rightarrow Target) = \frac{|\{X \subseteq T, T \in \mathscr{D}\}|}{|\mathscr{D}|} \times \tag{7.9}$$

$$\left(\frac{|\{X \cup Target \subseteq T, T \in \mathcal{D}\}|}{|\{X \subseteq T, T \in \mathcal{D}\}|} - \frac{|\{Target \subseteq T, T \in \mathcal{D}\}|}{|\mathcal{D}|} \right)$$

Specificity is another hybrid quality measure, which determines the proportion of negative cases incorrectly classified and it is computed as shown in Eq. (7.10).

$$specificity(X \rightarrow Target) = \frac{|\{\overline{X \cup Target} \subseteq T, T \in \mathcal{D}\}|}{|\{\overline{Target} \subseteq T, T \in \mathcal{D}\}|} \tag{7.10}$$

7.2.3 Deterministic Algorithms

Since the concept of subgroup discovery was introduced as a way of mining interesting relationships between variables in a set with respect to a specific target variable, this technique has been widely studied by many researchers and a number of algorithms have been proposed [29]. The first algorithms developed for the subgroup discovery task were defined under the names EXPLORA [34] and MIDOS [48]. These algorithms are defined as extensions of classification algorithms and they use decision trees to represent the knowledge.

Many other algorithms based on classification have been proposed for the subgroup discovery task [24, 35]. These algorithms look for individual rules of interest instead of generating models consisting of a set of rules inducing properties of all the classes of the target variable. Thus, different modifications of the original classification algorithms were implemented. For instance, SubgroupMiner [33] was proposed as an extension of both EXPLORA and MIDOS algorithms. The SubgroupMiner algorithm uses decision rules and an interactive search for discovering subgroups of interest in variables defined in discrete domains, so continuous variables require a previous discretization step. Finally, in order to quantify the interest of the discovery, several quality measures are used to verify whether the distribution of the target variable is statistically different in the extracted subgroup.

Gamberger et al. [24] proposed a subgroup discovery algorithm that aims to search for rules that maximize the precision quality measure (see Eq. (7.6)), which is a metric that measures the tradeoff of a subgroup between the number of transactions classified perfectly and the unusualness of their distribution. In this algorithm, high quality rules cover many transactions of the target variable and a low number of transactions of the non-target variable. The precision quality measure also includes a parameter g, which is responsible for determining the number of cases of the tolerated non-target variable relative to the number of covered cases of the target variable. Thus, low values of the parameter g implies rules with a high specificity, whereas high values imply more general rules, covering also transactions of the non-target variables. This algorithm helps the expert in performing flexible and effective searches on a wide range of optimal solutions instead of defining a fixed optimal measure to discover and automatically select the subgroups. The algorithm works in an iterative procedure by keeping in a pool the best subgroup in each iteration and replacing the worst subgroup in the pool if the new one is better.

CN2-SD [35] is another example of an approach based on classification algorithms and considered to be an adaptation of the CN2 classification algorithm [14]. CN2 works in an iterative fashion, searching for a conjunction of attributes that behaves well according to a heuristic measure, which is based on the information-theoretic entropy measure. This function prefers rules covering a large number of transactions of a single class and few of other classes. Then, the algorithm removes those transactions the extracted rule covers and adds the rule to a list of rules. The process iterates until no more satisfactory rules can be found. On the contrary, CN2-SD includes some modifications that can be summarized as follows: (1) it uses a new heuristic that combines generality (measured as the support of the rule) and accuracy (quantified as the confidence of the rule); (2) it also incorporates example weights into the covering step. Nevertheless, while in the original covering algorithm transactions are removed once they are covered by a rule, in the weighted covering algorithm transactions are never removed but their weight is decreased according some schemes; and (3) CN2-SD uses a probabilistic classification based on the target variable distribution of the transactions covered. A major drawback of CN2-SD is that it works on target variables with two values. In this regard, some authors [1] have proposed different algorithms for working with multi-class target values, i.e. CN2-MSD that is an extension of CN2-SD.

Many other authors [9, 32, 42] considered the subgroup discovery problem as an extension of association rule mining, which attempts to obtain relations between the variables in a dataset. In this case, different variables might appear both in the antecedent or consequent of the rule, the consequent meaning the target variable or property of interest. The characteristics of the association rule mining algorithms make it feasible to adapt these algorithms to the subgroup discovery task.

Apriori-SD [32] was proposed as modified version of Apriori-C [31], which an adaptation of the Apriori algorithm [3] for mining association rules. This adaptation to the subgroup discovery task includes a new post-processing mechanism, a new quality measure for the rules mined, and a probabilistic classification of the transactions. Based on Apriori-SD, Mueller et al. [42] proposed the SD4TS (Subgroup Discovery for Test Selection) algorithm. SD4TS was proposed as a new approach to test selection based on the discovery of subgroups of patients sharing the same optimal test, and present its application to breast cancer diagnosis. This algorithm was designed for finding the most cost-efficient subgroups of a population. It enables the output size to be reduced to the best t subgroups, pruning the search space significantly, especially for lower values of support. It should be noted that the algorithm was designed for a specific problem, so the quality measures used were particularly designed for this problem.

SD-Map [7] is another deterministic subgroup discovery algorithm defined as an extension of association rule mining. SD-Map uses the FP-Growth [28] algorithm, which differs from Apriori in its ability to reduce the database scans by considering a compressed representation of the database thanks to a data structure based on a traditional prefix-tree [46]. This structure represents each node by using an item and the frequency of the itemset denoted by the path from the root to that node. SD-Map adapts the model for mining association rules to the subgroup discovery task, considering a depth-first search step for evaluating the subgroups. Additionally, SD-Map uses a new procedure that can compute the subgroup quality directly without referring to other intermediate results.

7.2.4 Evolutionary Algorithms

Evolutionary algorithms [22] are mainly based on the principles of natural evolution and, more specifically, in the fact that organisms that are capable of acquiring resources will tend to have descendants in the future. Some authors [13] have considered that these algorithms are very suitable to perform the subgroups discovery task because they can reflect well the interaction of variables in rule-learning processes and provide wide flexibility in the representation [25].

Since subgroup discovery can be approached and solved as a search and optimization problem, some authors have treated the problem from an evolutionary perspective. Existing evolutionary algorithms in subgroup discovery are mainly based on a "chromosome = rule" approach and use a single objective as fitness function to be optimized. Hence, only one direction is considered to be optimized, which might include only one quality measure or a set of them that form an expression.

One of these algorithms is known as SDIGA [17], which is defined as an evolutionary model for mining fuzzy [49] rules for the subgroup discovery task. SDIGA has to be run once for each value in the target feature or class, so the consequent is not represented in the chromosome but is fixed. Therefore, each run of this algorithm mines a set of subgroups for a specific class or consequent value, and each subgroup is represented by means of fuzzy rules in disjunctive normal form, which offers a more flexible structure to the rules. The fuzzy sets corresponding to the linguistic labels are defined by means of the corresponding membership functions, which can be specified either by the user or by means of a uniform partition with triangular membership functions (see Fig. 7.2).

SDIGA represents each rule as a linear chromosome describing the values for the antecedent of the rule since all the individuals in the population are associated with the same value of the target feature. All the information relating to a rule is contained in a fixed-length chromosome with a binary representation where a single bit is used for each feasible value of an attribute (see Fig. 7.3). If the corresponding bit contains

Fig. 7.2 Example of fuzzy partition of a numerical variable considering three different linguistic labels

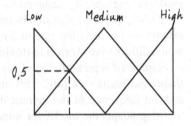

Fig. 7.3 Example of a rule encoded by SDIGA

0	1	0	0	1	1	0	0	0	0	1	0	0

Require: *maxGenerations* {Number of generations to be considered}
Ensure: *subgroups*
  1: *subgroups* ← / 0
  2: **for** ∀*value* ∈ *Target$_{feature}$* **do**
  3:   **repeat**
  4:     *rules* ← / 0
  5:     *number_generations* ← 0
  6:     create a set of *rules* whose consequent is *value*
  7:     **for** ∀*element* ∈ *rules* **do**
  8:        evaluate(*element*)
  9:     **end for**
 10:     **while** *number_generations* < *maxGenerations* **do**
 11:        *parents* ←select(*rules*)
 12:        *offspring* ←apply genetic operators to *parents*
 13:        **for** ∀*element* ∈ *offspring* **do**
 14:           evaluate(*element*)
 15:        **end for**
 16:        *R* ← take the best rule from {*rules, offspring*}
 17:        *rules* ← update(*rules, offspring, R*)
 18:        *number_generations* + +
 19:     **end while**
 20:     run a local search to improve *R*
 21:     **if** confidence(*R*)≥ minimum confidence and *R* represents new examples **then**
 22:        *subgroups* ← *subgroups* ∪ *R*
 23:        mark the set of transactions covered by *R*
 24:     **end if**
 25:   **until** confidence(*R*) < minimum confidence or *R* does not represent new examples
 26: **end for**

Fig. 7.4 Pseudo-code of the SDIGA algorithm

the value 1 or 0, it indicates that the value is used or not in the rule, respectively. In situations where a rule contains all the bits corresponding to an attribute with the value 1, or all of them contain the value 0, then the attribute has no relevance in the rule and it should be ignored (the attribute does not take part in the rule). Figure 7.3 illustrates an example rule comprising the second value for the first attribute and the first value for the fourth attribute. The second and the third attributes are ignored since all their bits are 0 or 1.

The pseudo-code of the SDIGA algorithm is shown in Fig. 7.4. The algorithm follows an iterative rule learning methodology (lines 3–25, Fig. 7.4) including a genetic algorithm to obtain the best rule for each specific value of a target feature (lines 5–19, Fig. 7.4). The genetic algorithm uses a steady-state reproduction model in which the original population is only modified through the substitution of the worst individuals by individuals resulting from crossover and mutation. After that, a local search is carried out as a postprocessing phase (line 20, Fig. 7.4) to increase the degree of support of the rule throughout a hill-climbing process.

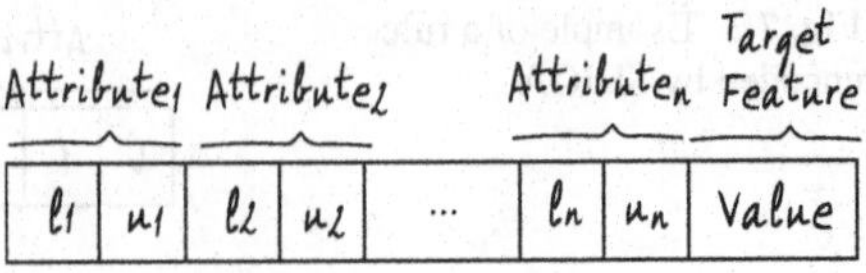

Fig. 7.5 Example of a rule
encoded by EDER-SD

	Attribute$_1$		Attribute$_2$		...		Attribute$_n$	Target Feature
l_1	u_1	l_2	u_2	...		l_n	u_n	Value

The aim of SDIGA is to extract understandable and general rules having a high accurate value. It means that the mining problem has at least two objectives to maximize: support (general rules) and confidence (accurate rules). Thus, the fitness function uses by SDIGA (see Eq. (7.11)) uses a weighted sum method that weights a set of objectives into a single expression, and the determination of proper values for the weights depends on the importance of each objective within the expression.

$$fitness(X \to Target) = \frac{w_1 \times support(X \to Target) + w_2 \times confidence(X \to Target)}{w_1 + w_2}$$

(7.11)

Another interesting evolutionary algorithm for subgroup discovery is EDER-SD [45] (Evolutionary Decision Rules for Subgroup Discovery). This evolutionary algorithm is capable of dealing with imbalanced data that generates rules only for the defective modules, the class of interest. EDER-SD represents each individual as a string of values where l_i and u_i represent the lower and upper interval for the i-th attribute, respectively. Finally, as shown in Fig. 7.5, the last position in the string represents the value of the target feature. In such situations where the lower l_i and upper u_i bounds of the interval of the attribute i-th are equal to their respective maximum boundaries, it means that the attribute is not relevant and will not appear in the rule.

EDER-SD follows an iterative rule learning methodology (see Fig. 7.6) so each iteration executes an evolutionary algorithm to induce a subgroup (the best discovered). Each transaction covered by a rule is penalised, hampering new rules that cover the same set of transactions (lines 20–24, Fig. 7.6). Therefore, during the first iteration of the evolutionary procedure, all transactions are equally considered, whereas in posteriori iterations, previously covered transactions are penalised and they are considered as less important for the fitness function calculation. Finally, it should be noted that EDER-SD is a single objective algorithm that includes a traditional genetic algorithm (lines 5–19, Fig. 7.6). Thus, the algorithm uses a single expression as a fitness function for the evolutionary process, and it allows the use of a wide number of the metrics used in subgroup discovery such as unusualness, confidence, sensitivity, or significance among others. Thus, rules are adapted to whatever measure is the most adequate in the domain.

Another approach based on an iterative rule learning methodology is GAR-SD [44]. GAR-SD is able to operate either with continuous (using intervals) and discrete attributes, representing rules as linear chromosomes that describe the values for the antecedent of the rule since all the individuals in the population are associated with the same value of the target feature. GAR-SD represents each individual as a string

Require: *maxGenerations* {Number of generations to be considered}
Ensure: *subgroups*
1: *subgroups* ← ∅
2: **for** ∀*value* ∈ *Target_feature* **do**
3:      **while** ∃ transaction for *value* ∈ *Target_feature* not covered **do**
4:          *rules* ← ∅
5:          *number_generations* ← 0
6:          create a set of *rules* whose consequent is *value*
7:          **for** ∀*element* ∈ *rules* **do**
8:              evaluate(*element*)
9:          **end for**
10:         **while** *number_generations* < *maxGenerations* **do**
11:             *parents* ←select(*rules*)
12:             *offspring* ←apply genetic operators to *parents*
13:             **for** ∀*element* ∈ *offspring* **do**
14:                 evaluate(*element*)
15:             **end for**
16:             *R* ← take the best rule from {*rules, offspring*}
17:             *rules* ← update(*rules, offspring, R*)
18:             *number_generations* + +
19:         **end while**
20:         **for** ∀*transaction* ∈ *data* **do**
21:             **if** *transaction* is covered by *R* **then**
22:                 penalise *transaction* using a threshold
23:             **end if**
24:         **end for**
25:         *subgroups* ← *subgroups* ∪ *R*
26:     **end while**
27: **end for**

Fig. 7.6 Pseudo-code of the EDER-SD algorithm

Fig. 7.7 Example of a rule
encoded by GAR-SD

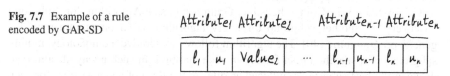

of values where each continuous attribute is represented as a tuple l_i and u_i, which represent the lower and upper intervals for the i-th attribute, respectively; whereas discrete attributes includes a single value (see second attribute in Fig. 7.7).

The GAR-SD algorithm (see Fig. 7.8) considers a weighted aggregation function comprising different objectives. Some of these objectives are related to some quality measures for subgroups discovery: support, confidence and significance. On the contrary, others are quality measures related to the variables that form the rule, such as the average amplitude of all the numeric intervals within the rule.

This algorithm is based on a traditional genetic algorithm (lines 5–19, Fig. 7.8), and it is designed to work with either continuous and discrete attributes. The aim of GAR-SD is to find the most interesting rules for each subgroup according to *Target_feature*. The algorithm uses a predefined threshold value (line 4, Fig. 7.8) to

Require: *maxGenerations* {Number of generations to be considered}
Ensure: *subgroups*
1: *subgroups* ← ∅
2: **for** ∀*value* ∈ *Target*$_{feature}$ **do**
3:      **while** number of non covered transactions < a specific threshold value **do**
4:          *rules* ← ∅
5:          *number_generations* ← 0
6:          create a set of *rules* whose consequent is *value*
7:          **for** ∀*element* ∈ *rules* **do**
8:              evaluate(*element*)
9:          **end for**
10:         **while** *number_generations* < *maxGenerations* **do**
11:             *parents* ←select(*rules*)
12:             *offspring* ←apply genetic operators to *parents*
13:             **for** ∀*element* ∈ *offspring* **do**
14:                 evaluate(*element*)
15:             **end for**
16:             *R* ← take the best rule from {*rules, offspring*}
17:             *rules* ← update(*rules, offspring, R*)
18:             *number_generations* + +
19:         **end while**
20:         check the set of non covered transactions
21:         *subgroups* ← *subgroups* ∪ *R*
22:     **end while**
23: **end for**

Fig. 7.8 Pseudo-code of the GAR-SD algorithm

indicate the percentage of examples covered by the rules of each target feature, which works as a stopping criterion.

CGBA-SD [39] (Comprehensible Grammar-Based Algorithm for Subgroup Discovery) is another subgroup discovery algorithm based on grammar guided genetic programming [40]. The algorithm follows an iterative methodology, mining subgroups of a specific target feature in each iteration. In such a way, its aim is to discover the best subgroups (induced as rules in the form of a syntax-tree) for each target feature or class. Individuals are encoded by means of a grammar (see Fig. 7.9) whose language is defined as $L(G) = \{(AND\ Condition)^n\ Condition \rightarrow Class : n \geq 0\}$. Therefore, using the aforementioned language, the grammar is well-defined and structured since any rule having at least one condition in the antecedent is obtained. Using this grammar it is possible to mine any subgroup containing either numerical or discrete features, which is an important feature of using grammars since the encoding of any individual can be adapted to each specific application domain or problem.

As mentioned above, CGBA-SD follows an iterative rule learning model, running a complete evolution of the rules for a specific value of a target feature iteratively, i.e., the individuals evolve over a number of generations in order to obtain the best ones. Once this generational procedure is finished, a new set of individuals satisfying a new value of the target feature is created. This iterative process is repeated as

$G = (\Sigma_N, \Sigma_T, P, S)$ with:

S = Subgroup
Σ_N = {Subgroup, Antecedent, Class, Nominal_Condition, Numerical_Condition }
Σ_T = {'AND', 'Attribute', 'Class_value', '=', 'IN','Min_value', 'Max_value', 'Target_attribute' }
P = {Subgroup ::= Antecedent, Class ;
      Antecedent ::= Nominal_Condition | Numerical_Condition |
                'AND', Nominal_Condition, Antecedent |
                'AND', Numerical_Condition, Antecedent ;
      Nominal_Condition ::= 'Attribute', '=', 'value' ;
      Numerical_Condition ::= 'Attribute', 'IN', 'Min_value', 'Max_value' ;
      Class ::= 'class', '=', 'Class_value'; }

Fig. 7.9 Context-free grammar used to encode individuals in the CGBA-SD algorithm

Require: *maxGenerations* {Number of generations to be considered}
Ensure: *subgroups*
1: **for** $\forall value \in Target_{feature}$ **do**
2:     *subgroups* $\leftarrow \emptyset$
3:     *rules* $\leftarrow \emptyset$
4:     *number_generations* $\leftarrow 0$
5:     create a set of *rules* whose consequent is *value*
6:     **for** $\forall element \in rules$ **do**
7:         evaluate(*element*)
8:     **end for**
9:     **while** *number_generations* < *maxGenerations* **do**
10:         *parents* $\leftarrow$ select(*rules*)
11:         *offspring* $\leftarrow$ apply genetic operators to *parents*
12:         **for** $\forall element \in offspring$ **do**
13:             evaluate(*element*)
14:         **end for**
15:         *rules* $\leftarrow$ update(*rules, offspring, subgroups*)
16:         *subgroups* $\leftarrow$ best individuals from {*rules, offspring, subgroups*} that overcome a minimum confidence threshold value
17:         *number_generations* $++$
18:     **end while**
19:     optimize bounds of the numerical intervals from *subgroups*
20:     return *subgroups*
21: **end for**

Fig. 7.10 Pseudo-code of the CGBA-SD algorithm

many times as the number of distinct values for the target features are available in the dataset.

In each generation of CGBA-SD (see Fig. 7.10), those rules (subgroups) exceeding a minimum confidence value previously specified are selected to be included in an external or elite population (see line 16, Fig. 7.10). This selection procedure works as an elitist selection, allowing some of the best individuals discovered from the current generation to be carried over unaltered to the next generation.

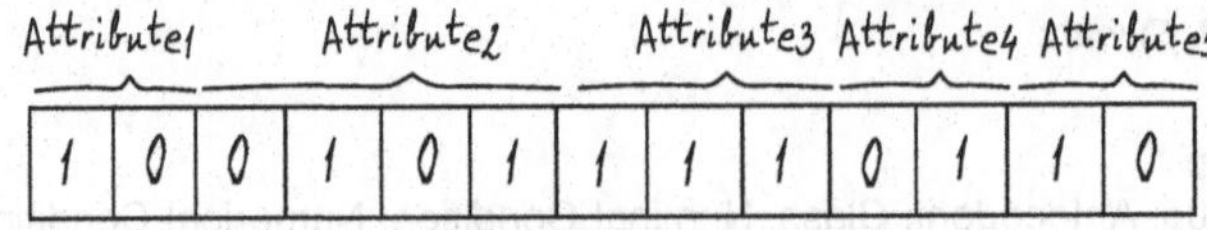

Attribute₁   Attribute₂   Attribute₃ Attribute₄ Attribute₅

| 1 | 0 | 0 | 1 | 0 | 1 | 1 | 1 | 1 | 0 | 1 | 1 | 0 |

Fig. 7.11 Example of a rule encoded by MESDIF, which determines that *Attribute₃* is meaningless since all its bits are equal to 1

The goal of the CGBA-SD algorithm is to mine reliable subgroups and cover a high percentage of examples of each target feature, so the better the fitness value (see Eq. (7.12)) the more promising the rule. A higher fitness function value enables measures of generality (considering the support based on examples of the class as shown in Eq. (7.2)) and precision (see Eq. (7.4)) to be optimized. Additionally, highly frequent subgroups imply some measures of complexity to be indirectly optimized, e.g. the number of variables in the antecedent of the rule will be smaller to obtain more frequent subgroups.

$$fitness(X \rightarrow Target) = support(X \rightarrow Target) \times confidence(X \rightarrow Target) \quad (7.12)$$

Finally, once the algorithm has discovered subgroups for each of the value of the target features, it runs a final procedure to optimize the intervals of the numerical conditions (see line 19, Fig. 7.10). The aim of this procedure is to increase or decrease the width of the intervals in order to discover high quality intervals that satisfy more instances or which prompt more reliable rules. This process is based on the well-known hill climbing optimization framework.

In subgroup discovery, the simultaneous optimization of different objectives at time is desirable in many situations, so different multi-objective evolutionary algorithms have been proposed to this end [11, 12]. Notice that many quality measures involve conflicting objectives so the improvement of one objective may lead to deterioration of another. In this situation, it does not exist a single solution that optimizes all objectives simultaneously and the best trade-off solutions are required. Typical multi-objective optimization approaches determine a set of solutions that is superior to the set when all the objectives are considered at time. Solutions of a specific iteration are organized in fronts [15] based on the objectives to be optimized and no solution of a front can be said to be better than the others within the same front. Nevertheless, solutions from the first front are better than solutions from the second front, and so on.

MESDIF (Multiobjective Evolutionary Subgroup DIscovery Fuzzy rules) [12] is a genetic algorithm based on a multi-objective methodology for mining fuzzy rules that represent subgroups. All the information relating to a rule is contained in a fixed-length chromosome (see Fig. 7.11) with a binary representation in which, for each attribute it is stored a bit for each of the possible values of the feature. If the corresponding bit contains the value 0 it indicates that the corresponding linguistic label or discrete value of the variable is not used in the rule, whereas the value 1 indicates that the corresponding value is included. It is interesting to note that such

Require: *population_size, elite_size*{Population and elite population sizes}
Ensure: *subgroups*
1: *subgroups* ← ∅
2: **for** ∀*value* ∈ *Target_feature* **do**
3: *population* ← ∅
4: *elite_population* ← ∅
5: *parents* ← ∅
6: *offspring* ← ∅
7: *population* ← generate(*population_size*)
8: **for** ∀*member* ∈ *population* **do**
9:   calculate the fitness value of *member*
10: **end for**
11: **while** termination condition is not reached **do**
12:   *elite_population* ← obtain the pareto front from {*population, elite_population*}
13:   **if** size(*pareto_front*) > *pareto_size* **then**
14:    *elite_population* ← reduce the set *elite_population* to the size of *elite_size*
15:   **else**
16:    **if** size(*elite_population*) < *elite_size* **then**
17:     *elite_population* ← add new solutions to *elite_population* from *population*
18:    **end if**
19:   **end if**
20:   *parents* ← obtain a set of individuals from *population* by a binary tournament selector
21:   *offspring* ← applyGeneticOperators(*parents*)
22:   **for** ∀*member* ∈ *population* **do**
23:    calculate the fitness value of *member*
24:   **end for**
25:   *population* ← update(*population, offspring*)
26: **end while**
27: *subgroups* ← *subgroups* ∪ *elite_population*
28: **end for**
29: return *subgroups*

Fig. 7.12 Pseudo-code of the MESDIF algorithm

attributes comprising all the bits with the value 1 indicate that they have no relevance in the rule since all the values verify the rule condition.

The search process of MESDIF is inspired by SPEA2 [52] (see Fig. 7.12), which is one of the most well-known multi-objective algorithms. MESDIF uses several quality measures (confidence, support, significance and unusualness) at the same time to evaluate the quality of the subgroups, and applies the concepts of elitism in the selection of the best rules or subgroups by using a secondary or elite population (lines 14–21, Fig. 7.12). In order to preserve the diversity of the set of solutions, MESDIF considers a niches technique that analyses the proximity in values of the objectives and an additional objective based on novelty to promote rules which give information on examples not described by other rules of the population.

Another multi-objective evolutionary algorithm for mining fuzzy rules for the subgroup discovery task was proposed by Carmona et al. [11]. This algorithm, named NMEEF-SD (Non-dominated Multi-objective Evolutionary algorithm for Extracting Fuzzy rules in Subgroup Discovery), is based on a multi-objective

Fig. 7.13 Example of a rule
encoded by NMEEF-SD

Attr.$_1$	Attr.$_2$	Attr.$_3$	Attr.$_4$	Attr.$_5$
2	1	4	2	0

approach to optimize a set of quality measures at time (confidence, support, sensitivity, significance and unusualness). The search strategy of NMEEF-SD is similar to that of NSGA-II [16], which returns a predefined number of optimal solutions previously organized in fronts.

NMEEF-SD permits the representation of solutions including numerical attributes without the need for a previous discretization. In this sense, this algorithm uses fuzzy rules, which contribute to the interpretability of the extracted rules since it uses a knowledge representation close to the expert. Thus, similarly to the previously described MESDIF algorithm [12], NMEEF-SD uses fuzzy logic to represent the continuous variables by means of linguistic variables that allow numerical attributes to be used without the need to increase the interpretability of the extracted knowledge by their discretization. The continuous variable are then considered linguistic, and the fuzzy sets corresponding to the linguistic labels can be specified by the user or defined by means of a uniform partition if the expert knowledge is not available. NMEEF-SD uses an integer representation (see Fig. 7.13) model with as many genes as variables contained in the original dataset without considering the target variable. The set of possible values for the categorical features is that indicated by the problem, and for numerical variables, it is the set of linguistic terms determined heuristically or with expert information. Thus, for a dataset comprising five attributes, Fig. 7.13 illustrates a sample rule where the first attribute takes the second discrete value (or second linguistic label), the second attribute takes the first discrete value (or first linguistic label), and so on. It should be noted that a zero value is used to indicate that the variable is not considered by the rule.

NMEEF-SD is based on the NSGA-II algorithm [16], and it evolves the set of individuals that are organized in fronts [15] based on the objectives to be optimized. No solution of a front can be said to be better than the others within the same front, but solutions from the first front are better than solutions from the second front, and so on. NMEEF-SD generate individuals considering a biased initialization (see line 9, Fig. 7.14) whose aim is to obtain an initial population with general individuals that cover a high number of transactions. This operator generates some of the individuals of the population using only a maximum percentage of variables for each rule, so the algorithm begins with a set of rules with high generality because most of the generated individuals are rules with a low number of variables.

Another interesting procedure proposed by NMEEF-SD is the re-initialization based on coverage (see line 26, Fig. 7.14). The last step for each generation is to check whether *pareto_front* is evolving or not. It is considered that *pareto_front* evolves if it covers at least one example of the dataset not covered by *pareto_front* of the previous generation. If *pareto_front* does not evolve for more than 5 % of the evolutionary process (quantified through the number of evaluations) re-initialization

Require: *population_size, elite_size* {Population and elite population sizes}
Ensure: *subgroups*
 1: *subgroups* ← ∅
 2: **for** ∀*value* ∈ *Target_feature* **do**
 3:      *population* ← ∅
 4:      *pareto_front* ← ∅
 5:      *parents* ← ∅
 6:      *offspring* ← ∅
 7:      *population* ← generate *population_size* with biased initialization
 8:      **while** termination condition is not reached **do**
 9:          *parents* ← obtain a set of individuals from *population* by a tournament selector
10:          *offspring* ← apply genetic operators on the set *parents*
11:          **for** ∀*member* ∈ *population* **do**
12:              evaluate(*member*)
13:          **end for**
14:          *population* ← update(*population, offspring*)
15:          generate all non-dominated fronts $F = (F_1, F_2, ...)$ from population and calculate crowding-distance in F_i
16:          *pareto_front* ← F_1
17:          **if** *pareto_front* evolves **then**
18:              *i* ← 2
19:              **while** $|population| + |F_i| \leq population\_size$ **do**
20:                  *population* ← *population* ∪ F_i
21:                  *i*++
22:              **end while**
23:          **else**
24:              apply re-initialization based on coverage
25:          **end if**
26:      **end while**
27: **end for**
28: return *subgroups*

Fig. 7.14 Pseudo-code of the NMEEF-SD algorithm

is performed through the following steps: (1) the non-repeated individuals of *pareto_front* are directly replicated in the new population; (2) the new population is completed with individuals generated through initialization based on coverage.

7.3 Other Supervised Local Pattern Mining Approaches

The task of mining supervised local patterns is related to the finding of patterns that describe subsets of data with a high statistical unusualness in their distribution, and these subsets are formed by using a single target feature or a set of them. Different approaches for supervised local pattern mining have been proposed, which mainly differ in the way in which the statistical unusualness is quantified. Contrast set mining [43] is one of these approaches, and it is closely related to association rule mining [3]. Contrast sets were first defined as conjunctions of attributes and values that differ meaningfully in their distributions across groups. Discovering the differences between groups might be essential in many fields. For example, in medical domains, it might be interesting to determine the differences between males and females affected by diabetes. In this regard, it is possible to discover that women who have a waist size greater than 40 inches, are twice as likely to have diabetes, than men who have the same waist size. Contrast set mining seeks conjunctions of attributes and values, called contrast sets, whose frequencies of occurrence are different in different groups.

STUCCO (Search and Testing for Understandable Consistent Contrasts) is one of the first algorithms [8] for the discovery of contrast sets along with their supports on groups. The support of a contrast set with respect to a group G_i, represented as $support(CS, G_i)$, is the percentage of transactions in G_i satisfied by the contrast set CS. STUCCO specifically finds significant contrast sets by calculating $max_{ij}(|support(CS, G_i)support(CS, G_j)|) \geq \delta$, where δ is a user defined threshold called the minimum support difference. In the contrast set mining process, STUCCO represents the search space for potential contrast sets as a tree structure having every possible combination of the attributes. The root of the tree represents the entire dataset, whereas subsequent levels will represent conjunctions of different attributes (two attributes for the second level, three for the third, and so on). The number of levels in the search tree is equal to the number of attributes, and the number of contrast sets on each level is equal to C_i^n, where n is the number of attributes and i is the level of the tree. It should be noted that the set $\{a, b\}$ represents the same as $\{b, a\}$. Thus, the total number of contrast sets that is possible to obtain is equal to $\sum_{i=1}^{n} C_i^n$.

In the literature, there are new and different approaches for mining contrast sets [30, 38], and some of them were developed for discovering contrasts that can include negations of terms in the contrast set [47]. Finally, it is interesting to note that most of the approaches for contrast set mining generate all candidate contrast sets from discrete (or discretized) data and later use statistical tests to identify the interesting ones.

A different approach for supervised local pattern mining is known as emerging pattern mining. Emerging patterns are defined as itemsets whose support increases significantly from one data set to another [18]. Emerging patterns are said to capture emerging trends in time-stamped databases, or to capture differentiating characteristics between classes of data. From a semantic point of view, emerging patterns are association rules with an itemset in rule antecedent, and a fixed consequent: $ItemSet \rightarrow D_1$, for given data set D_1 being compared to another data set D_2. The measure of quality of emerging patterns is the growth rate, which is obtained by comparing the support of the rule in D_1 with regard to the support of the rule in D_2. As a matter of example, let us considered a pattern that is satisfied in 10 % of the transactions in a dataset D_1 and it satisfied 5 % of the transactions for a different dataset D_2. This pattern is said to be better than another with a support value of 70 % in one data set and 50 % in the other $\left(\dfrac{10}{5} > \dfrac{70}{50}\right)$.

The main problem when mining emerging patterns is that the set of patterns discovered may be too big, even when some minimum growth rate constraints are used, so the analysis of such huge set of patterns by an expert is not an easy task. In this sense, some researchers have focused on selecting the interesting emerging patterns and representing a condensed set of patterns [23]. Focusing on the applications field, emerging patterns have been mainly applied to bioinformatics, and more specifically, to microarray data analysis [37] for finding groups of genes by emerging patterns and applied it to the ALL/AML data set and different tumor databases.

References

1. T. Abudawood and P. Flach. Evaluation measures for multi-class subgroup discovery. In W. Buntine, M. Grobelnik, D. Mladenić, and J. Shawe-Taylor, editors, *Machine Learning and Knowledge Discovery in Databases*, volume 5781 of *Lecture Notes in Computer Science*, pages 35–50. Springer Berlin Heidelberg, 2009.
2. C. C. Aggarwal and J. Han. *Frequent Pattern Mining*. Springer International Publishing, 2014.
3. R. Agrawal, T. Imielinski, and A. N. Swami. Mining association rules between sets of items in large databases. In *Proceedings of the 1993 ACM SIGMOD International Conference on Management of Data*, SIGMOD Conference '93, pages 207–216, Washington, DC, USA, 1993.
4. J. Alípio, F. Pereira, and P. J. Azevedo. Visual interactive subgroup discovery with numerical properties of interest. In L. Todorovski, N. Lavrač, and K. Jantke, editors, *Discovery Science*, volume 4265 of *Lecture Notes in Computer Science*, pages 301–305. Springer Berlin Heidelberg, 2006.
5. M. L. Antonie and O. R. Zaïane. Text Document Categorization by Term Association. In *Proceedings of the 2002 IEEE International Conference on Data Mining*, ICDM '02, pages 19–26, Washington, DC, USA, 2002. IEEE Computer Society.
6. M. Atzmueller. Subgroup Discovery - Advanced Review. *WIREs: Data Mining and Knowledge Discovery*, 5:35–49, 2015.
7. M. Atzmueller and F. Puppe. SD-Map – A Fast Algorithm for Exhaustive Subgroup Discovery. In *Proceedings of the 10th European Symposium on Principles of Data Mining and Knowledge Discovery*, PKDD '06, pages 6–17, Berlin, Germany, 2006.
8. S. D. Bay and M. J. Pazzani. Detecting Group Differences: Mining Contrast Sets. *Data Mining and Knowledge Discovery*, 5(3):213–246, 2001.
9. M. Boley and H. Grosskreutz. Non-redundant subgroup discovery using a closure system. In *Proceedings of the 2009 European Conference on Machine Learning and Principles and Practice of Knowledge Discovery in Databases*, ECML/PKDD 2009, pages 179–194, Bled, Slovenia, September 2009. Springer.
10. O. Bousquet, U. Luxburg, and G. Ratsch. *Advanced Lectures On Machine Learning*. SpringerVerlag, 2004.
11. C. J. Carmona, P. González, M. J. del Jesus, and F. Herrera. NMEEF-SD: Non-dominated multiobjective evolutionary algorithm for extracting fuzzy rules in subgroup discovery. *IEEE Transactions on Fuzzy Systems*, 18(5):958–970, 2010.
12. C. J. Carmona, P. González, M. J. del Jesus, M. Navío-Acosta, and L. Jiménez-Trevino. Evolutionary fuzzy rule extraction for subgroup discovery in a psychiatric emergency department. *Soft Computing*, 15(12):2435–2448, 2011.
13. C. J. Carmona, P. González, M. J. del Jesus, and F. Herrera. Overview on evolutionary subgroup discovery: analysis of the suitability and potential of the search performed by evolutionary algorithms. *Wiley Interdisciplinary Reviews: Data Mining and Knowledge Discovery*, 4(2): 87–103, 2014.

14. P. Clark and T. Niblett. The cn2 induction algorithm. *Machine Learning*, 3(4):261–283, 1989.
15. C. A. Coello, G. B. Lamont, and D. A. Van Veldhuizen. *Evolutionary Algorithms for Solving Multi-Objective Problems (Genetic and Evolutionary Computation)*. Springer-Verlag New York, Inc., Secaucus, NJ, USA, 2006.
16. K. Deb, A. Pratap, S. Agrawal, and T. Meyarivan. A Fast Elitist Multi-Objective Genetic Algorithm: NSGA-II. *IEEE Transactions on Evolutionary Computation*, 6:182–197, 2000.
17. M. J. del Jesus, P. González, F. Herrera, and M. Mesonero. Evolutionary fuzzy rule induction process for subgroup discovery: A case study in marketing. *IEEE Transactions on Fuzzy Systems*, 15(4):578–592, 2007.
18. G. Dong and J. Li. Efficient mining of emerging patterns: Discovering trends and differences. In *Proceedings of the 5th ACM SIGKDD International Conference on Knowledge Discovery and Data Mining*, KDD '99, pages 43–52, New York, NY, USA, 1999.
19. G. Dong and J. Li. Emerging patterns. In L. Liu and M. T. Özsu, editors, *Encyclopedia of Database Systems*, pages 985–988. Springer US, 2009.
20. W. Duivesteijn and A. J. Knobbe. Exploiting false discoveries - statistical validation of patterns and quality measures in subgroup discovery. In *Proceedings of the 11th IEEE International Conference on Data Mining*, ICDM 2011, pages 151–160, Vacouver, BC, Canada, December 2011.
21. W. Duivesteijn, A. J. Knobbe, A. Feelders, and M. van Leeuwen. Subgroup discovery meets Bayesian networks – an exceptional model mining approach. In *Proceedings of the 2010 IEEE International Conference on Data Mining*, ICDM 2010, pages 158–167, Sydney, Australia, December 2010. IEEE Computer Society.
22. D. Dumitrescu, B. Lazzerini, L. C. Jain, and A. Dumitrescu. *Evolutionary Computation*. CRC Press, Inc., Boca Raton, FL, USA, 2000.
23. H. Fan and K. Ramamohanarao. Efficiently mining interesting emerging patterns. In G. Dong, C. Tang, and W. Wang, editors, *Advances in Web-Age Information Management*, pages 189–201. Springer Berlin Heidelberg, 2003.
24. ssss D. Gamberger and N. Lavrac. Expert-guided subgroup discovery: Methodology and application. *Journal of Artificial Intelligence Research*, 17:501–527, 2002.
25. P. González-Espejo, S. Ventura, and F. Herrera. A Survey on the Application of Genetic Programming to Classification. *IEEE Transactions on Systems, Man and Cybernetics: Part C*, 40(2):121–144, 2010.
26. H. Grosskreutz and S. Ruping. On subgroup discovery in numerical domains. *Data Mining and Knowledge Discovery*, 19(2):210–226, 2009.
27. J. Han and M. Kamber. *Data Mining: Concepts and Techniques*. Morgan Kaufmann, 2000.
28. J. Han, J. Pei, Y. Yin, and R. Mao. Mining Frequent Patterns without Candidate Generation: A Frequent-Pattern Tree Approach. *Data Mining and Knowledge Discovery*, 8:53–87, 2004.
29. F. Herrera, C. J. Carmona, P. González, and M. J. del Jesus. An overview on subgroup discovery: Foundations and applications. *Knowledge and Information Systems*, 29(3):495–525, 2011.
30. R. J. Hilderman and T. Peckham. A statistically sound alternative approach to mining contrast sets. In *Proceedings of the 4th Australasian Data Mining Conference*, AusDM 2005, pages 157–172, Sydney, Australia, 2005.
31. Viktor Jovanoski and Nada Lavrač. Classification rule learning with APRIORI-C. In *Proceedings of the 10th Portuguese Conference on Artificial Intelligence on Progress in Artificial Intelligence, Knowledge Extraction, Multi-agent Systems, Logic Programming and Constraint Solving*, EPIA '01, pages 44–51, London, UK, 2001. Springer-Verlag.
32. B. Kavsek and N. Lavrač. APRIORI-SD: Adapting association rule learning to subgroup discovery. *Applied Artificial Intelligence*, 20(7):543–583, 2006.
33. W. Kloesgen and M. May. Census data mining an application. In *In Proceedings of the 6th European Conference on Principles and Practice of Knowledge Discovery in Databases*, PKDD 2002, pages 733–739, Helsinki, Finland, 2002. Springer-Verlag London.

34. W. Klösgen. Explora: A multipattern and multistrategy discovery assistant. In U. M. Fayyad, G. Piatetsky-Shapiro, P. Smyth, and R. Uthurusamy, editors, *Advances in Knowledge Discovery and Data Mining*, pages 249–271. American Association for Artificial Intelligence, 1996.
35. N. Lavrač, B. Kavšek, P. Flach, and L. Todorovski. Subgroup discovery with cn2-sd. *Journal of Machine Learning Research*, 5:153–188, December 2004.
36. D. Leman, A. Feelders, and A. J. Knobbe. Exceptional model mining. In *Proceedings of the European Conference in Machine Learning and Knowledge Discovery in Databases*, volume 5212 of *ECML/PKDD 2008*, pages 1–16, Antwerp, Belgium, 2008. Springer.
37. J. Li and L. Wong. Identifying good diagnostic gene groups from gene expression profiles using the concept of emerging patterns. *Bioinformatics*, 18(5):725–734, 2002.
38. J. Lin and E. J. Keogh. Extending the notion of contrast sets to time series and multimedia data. In *Proceedings of the 10th European Conference on Principles and Practice of Knowledge Discovery in Databases*, PKDD 2006, pages 284–296, Berlin, Germany, 2006.
39. J. M. Luna, J. R. Romero, C. Romero, and S. Ventura. On the use of genetic programming for mining comprehensible rules in subgroup discovery. *IEEE Transactions on Cybernetics*, 44(12):2329–2341, 2014.
40. R. McKay, N. Hoai, P. Whigham, Y. Shan, and M. O'Neill. Grammar-based Genetic Programming: a Survey. *Genetic Programming and Evolvable Machines*, 11:365–396, 2010.
41. K. Moreland and K. Truemper. Discretization of target attributes for subgroup discovery. In *Proceedings of the 6th International Conference on Machine Learning and Data Mining in Pattern Recognition*, MLDM 2009, pages 44–52, Leipzig, Germany, 2009. Springer.
42. M. Mueller, R. Rosales, H. Steck, S. Krishnan, B. Rao, and S. Kramer. Subgroup discovery for test selection: A novel approach and its application to breast cancer diagnosis. In N. Adams, C. Robardet, A. Siebes, and J. F. Boulicaut, editors, *Advances in Intelligent Data Analysis VIII*, volume 5772 of *Lecture Notes in Computer Science*, pages 119–130. Springer Berlin Heidelberg, 2009.
43. P. K. Novak, N. Lavrač, and G. I. Webb. Supervised descriptive rule discovery: A unifying survey of contrast set, emerging pattern and subgroup mining. *Journal of Machine Learning Research*, 10:377–403, 2009.
44. V. Pachón, J. Mata, J. L. Domínguez, and M. J. Maña. A multi-objective evolutionary approach for subgroup discovery. In *Proceedings of the 5th International Conference on Hybrid Artificial Intelligence Systems*, HAIS 2010, pages 271–278, San Sebastian, Spain, 2010. Springer.
45. D. Rodriguez, R. Ruiz, J. C. Riquelme, and J. S. Aguilar-Ruiz. Searching for rules to detect defective modules: A subgroup discovery approach. *Information Sciences*, 191:14–30, 2012.
46. P. N. Tan, M. Steinbach, and V. Kumar. *Introduction to Data Mining*. Addison Wesley, 2005.
47. T. T. Wong and K. L. Tseng. Mining negative contrast sets from data with discrete attributes. *Expert Systems with Applications*, 29(2):401–407, 2005.
48. S. Wrobel. An algorithm for multi-relational discovery of subgroups. In *Proceedings of the 1st European Symposium on Principles of Data Mining and Knowledge Discovery*, PKDD '97, pages 78–87, London, UK, UK, 1997. Springer-Verlag.
49. L. A. Zadeh. The concept of a linguistic variable and its application to approximate reasoning I,II,III. *Information Sciences*, 8–9:199–249, 301–357, 43–80, 1975.
50. A. Zimmermann and S. Nijssen. Supervised pattern mining and applications to classification. In C. C. Aggarwal and J. Han, editors, *Frequent Pattern Mining*, pages 425–442. Springer International Publishing, 2014.
51. A. Zimmermann, B. Bringmann, and R. Ulrich. Fast, effective molecular feature mining by local optimization. In *Proceedings of the 2010 European Conference on Machine Learning and Principles and Practice of Knowledge Discovery in Databases*, ECML/PKDD 2010, pages 563–578, Barcelona, Spain, 2010. Springer.
52. E. Zitzler, M. Laumanns, and L. Thiele. SPEA2: Improving the Strength Pareto Evolutionary Algorithm for Multiobjective Optimization. In *Proceedings of the 2001 conference on Evolutionary Methods for Design, Optimisation and Control with Application to Industrial Problems*, EUROGEN 2001, pages 95–100, Athens, Greece, 2001.

Chapter 8
Mining Exceptional Relationships Between Patterns

Abstract In any dataset, it is possible to identify small subsets of data which distribution is exceptionally different from the distribution in the complete set of data records. Finding such exceptional behaviour it is possible to mine interesting associations, which is known as the exceptional relationship mining task. This chapter formally describes the task, which is considered as a special process that lies in the intersection of both exceptional model mining and association rule mining. The current chapter first analyses the exceptional model mining task. Then, it formally describes the task of mining exceptional relationships. Finally, this chapter deals with a model based on grammars and some applications where the extraction of exceptional relationships is justified.

8.1 Introduction

By and large, the extraction of patterns and association between patterns, as well as their description and analysis ease the understanding of the information gathered in multiple domains. Patterns represent any type of homogeneity and regularity in data, describing intrinsic and important hidden properties [1]. The extraction of these properties, however, does not encompass all forms of knowledge so it may not be enough for some application domains where it is interesting to discover exceptional models—groups of patterns whose distribution is exceptionally different from that of the entire data [12].

The task of discovering exceptional models is the task of identifying elements that behave exceptionally from the norm in data. Most of the data mining researches in this field are focused on detecting outliers [9, 13]. However, the mining of exceptional patterns in local pattern mining [10] is related to the looking for any deviation in a subset of records instead of in the whole data. Several target features are selected, and a model over these features is chosen to be the target concept [5]. Then, the aim is to discover subgroups [11] represented by single features where the model over the targets on the subgroup is substantially different from the model on the whole dataset.

Exceptional model mining was proposed as an interesting task that provides new properties from data. As a matter of example, let us consider a dataset with information about people of different countries around the world. When

© Springer International Publishing Switzerland 2016

S. Ventura, J.M. Luna, *Pattern Mining with Evolutionary Algorithms*,

DOI 10.1007/978-3-319-33858-3_8

analysing the relationship between age and mortality rate, it is obtained that both are positively correlated, i.e. the higher the age, the higher the mortality rate. However, when analysing the same relationship by including the pattern {*poverty, infectious_disease*}, it is discovered that both age and mortality rate are not positively correlated. In fact, it is discovered that the first year of life is crucial in the survival of human beings and it is high probable to die at an early age, i.e. the correlation is negative now.

Exceptional models describe useful information not provided by patterns in isolation. The extraction and analysis of this kind of models is cornerstone in many application fields, providing richer information about the data behaviour. These models are defined as sets of items that guarantee the relative frequency of the subgroups in the database. Nevertheless, similarly to frequent pattern mining, these subgroups are considered as a whole and do not describe relations between the items. At this point, the extraction of associations might provide interesting features that together with the extracted exceptional models give rise to exceptional relationship mining. This task is considered as a special case that lies in the intersection of both exceptional model mining and association rule mining [22]. Considering the aforementioned exceptional pattern {*poverty, infectious_disease*}, it is possible to identify strong associations defined as exceptional relationships. For example, the exceptional rule *poverty → infectious_disease* describes that it is highly probable that someone that is poor tends to get an infectious disease. This rule is much more accurate than the rule *infectious_disease → poverty*, since poor people is more vulnerable to infectious diseases, whereas a infectious diseases does not always imply poverty.

The mining of exceptional relationships enables the description of reliable, abnormal and exceptional relations between items to be discovered so its computational cost in the mining process is much higher than the one of exceptional models, requiring the mining of both exceptional subgroups and association rules within each exceptional subgroup. Considering the computational cost as a real drawback of the exceptional relationship mining task, the use of grammars to include syntax constraints is a really interesting way to reduce the search cost. Additionally, grammars can be considered as a useful way of introducing subjective and external knowledge to the pattern mining process, enabling patterns of interest to be obtained by users with different goals. Grammars are highly related to the background knowledge of the user [8], and they allow to define solutions that have flexible and expressive structures that ease the interpretability of the extracted knowledge. All of this cause the definition of the first model for mining exceptional relationships by means of a grammar-guided genetic programming model [19], which achieved excellent results in unsupervised learning tasks [14].

8.2 Mining the Exceptionableness

This section formally describes the concept of exceptionalness. In this regard, it first introduces the exceptional model mining problem, which is considered as an extension of subgroup discovery [3]. Then, this section formally describes the problem of mining of exceptional relationships [16] among patterns, which is considered as a task defined halfway between mining association rules and exceptional models.

8.2.1 *Exceptional Model Mining Problem*

Leman et al. [12] formally described the exceptional model mining problem as an extension of subgroup discovery [10], so the aim is to discover subgroups where a model fitted to the subgroup is substantially different from the same model fitted to the entire database [4]. Let us assume that the database $\mathscr{D}$ is a set of transactions $t \in \mathscr{D}$, and each description of a transaction includes a number of features $f_1, \ldots, f_k$. A pattern P in a database $\mathscr{D}$ is defined as a function $P \rightarrow \{0, 1\}$ so that the function will take the value 1, i.e. $P(t_i) = 1$, for the i-th transaction if and only if P satisfies t_i. It is said that the pattern P satisfies the transaction t_i if and only if P is a subset of features $f_i, \ldots, f_j$ of t_i. Additionally, a subgroup G_P for a pattern P is formally defined as the set of transactions $G_P \subseteq \mathscr{D}$ that are satisfied by P, i.e. $G_P = \{\forall t_i \in \mathscr{D} : P(t_i) = 1\}$. The size of a subgroup G_P is denoted as $|G_P|$ and determines the number of transactions satisfied by P. As for the complement of a subgroup $\overline{G}_P$, it is defined as the set of transactions $\overline{G}_P \subseteq \mathscr{D}$ that are not satisfied by P, i.e. $\overline{G}_P = \mathscr{D} \setminus G_P$.

In exceptional model mining, it is required a set of conditions, also known as target features, where a subgroup G_P presents a specific behaviour. In each subgroup defined on a set of target features, it is obtained a model that describes the behaviour of G_P. In order to quantify how exceptional del obtained model is in relation to the model induced on its complement, some authors have considered a set of quality measures [21]. In such situations where only the subgroup itself is considered, the quality measures tend to focus on the accuracy of the model, such as the fit of a regression line. If the quality measure captures the difference between a subgroup G_P and its complement $\overline{G}_p$, it is typically based on a comparison between more structural properties of the two models, such as the slope of the regression lines.

As a matter of example, let us consider again the sample dataset with information about people of different countries around the world. Features *age* and *mortality rate* are considered as target features where the model is represented. As shown in Fig. 8.1, in a general scenario, the regression line determines that the higher the age, the higher the mortality rate in different countries. Based on this example, it is possible to describe a pattern $P = \{poverty, infectious\_disease\}$, so the distribution of the target features can be divided into G_P and $\overline{G}_P$ as it is graphically

Fig. 8.1 Distribution and
regression line of two target
features (*age* and *mortality
rate*) from information of
different countries around the
world

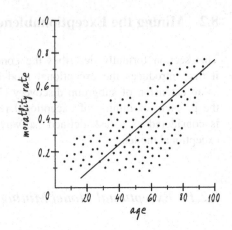

shown in Fig. 8.2. G_P represents a subgroup of transactions that describes people
who live in poverty and suffer from an infectious disease. It should be noted that
most infectious diseases are caused by an inefficient sanitary system and it is more
likely to get sick if your immune system is not working properly due to disorders
that affect your immune system, i.e. malnutrition. Additionally, severe poverty is
the root cause of the high mortality rates in the world and these rates are higher
at early ages in underdeveloped countries. Analysing the distribution of G_P with
regard to its complement $\overline{G}_P$ (see Fig. 8.2), it is obtained that their regression lines
are quite different, having a different slope. In fact, G_P does not present a higher
rate of mortality when the age increases, so this behaviour can be considered as
exceptional.

Considering the concept of exceptional model mining, it can be formally defined
as given a dataset comprising a set of transactions $\mathcal{D}$, and a measure ϕ that quantifies
the exceptionalness according to a minimum threshold ϵ. The task is to find all
subgroups $\mathcal{G}$ with corresponding minimal subgroup description, such that each $G \in
\mathcal{G}$ implies an exceptional model, i.e., $\phi^{\mathcal{G}}(G) \geq \epsilon$.

Analysing the task of discovering exceptional models, it may be related to the
Simpson's paradox [17], which is a contradiction that may be found in probability
and statistics. The paradox occurs when a specific trend appears in different datasets
but disappears when all the datasets are combined:

$$\frac{A}{B} > \frac{a}{b}, \frac{C}{D} > \frac{c}{d} \not\Rightarrow \frac{A+C}{B+D} > \frac{a+c}{b+d}$$

A well-known real-world example of the *Simpson's* paradox is described in [7].
When analysing different applications for admission at the university of California,
Berkeley, it was discovered that men were more likely than women to be admitted,
and the difference was so large that some remedial actions were required. For
instance, 44 % of men were admitted, whereas only 35 % of women do. Nevertheless, when examining individual departments, it was discovered that no department

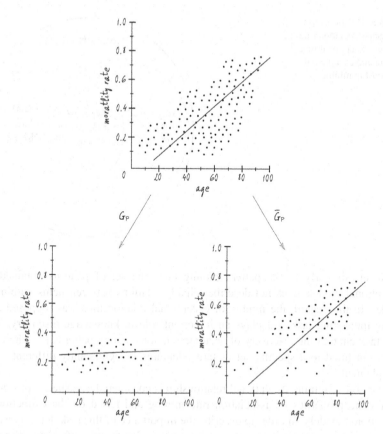

Fig. 8.2 Distribution of the target features *age* and *mortality rate* for the whole dataset and where G_P and $\overline{G}_P$ comes to play. It should be noted that the patten P is defined as $P = \{poverty, infectious\_disease\}$ for this example

was significantly biased against women and some of them had small but statistically significant bias in favor of women, i.e. 82 % of women were admitted to a specific department whereas only 62 % of men do.

8.2.2 Exceptional Relationship Mining

The mining of exceptional relationships [16] among patterns was proposed as a task described halfway between association rule mining and exceptional model mining (see Fig. 8.3). Focusing on exceptional models [12], they describe useful information not provided by many other descriptive techniques like pattern mining. In many application fields, the information provided by exceptional models is cornerstone and richer information about the data behaviour can be obtained.

Fig. 8.3 Relationship
between association rule
mining, exceptional model
mining and exceptional
relationship mining

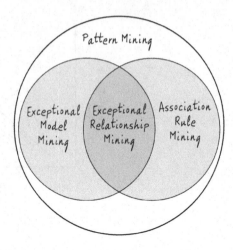

However, similarly to the pattern mining task, the set of patterns obtained by exceptional models does not describe reliable relations between items within the model. In this regard, the mining of exceptional relationships was proposed as a really interesting task that plays an important role in knowledge discovery. This new task enables the discovery of accurate relations among patterns [2] where the subgroup fitted to a specific set of target features is significantly different to its complement.

The task of mining exceptional relationship is considered as a special procedure with some properties of association rule mining and based on the extraction of exceptional models. In order to describe the importance of this task, let us consider again the aforementioned exceptional pattern $P = \{poverty, \ infectious\_disease\}$. This pattern P, obtained thanks to the mining of exceptional models, identifies only a exceptional behaviour in data so it does not provide any descriptive behaviour among its items. In such a way, association rule mining can provide a better description of the extracted exceptional model, being able to identify strong associations within the model. As a matter of example, let us consider the exceptional rule $poverty \rightarrow infectious\_disease$ obtained from the exceptional pattern $P = \{poverty, \ infectious\_disease\}$. This rule describes that it is highly probable that someone that is poor tends to get an infectious disease. Similarly, it is also possible to obtain the rule $infectious\_disease \rightarrow poverty$, which is less accurate than the previous one since poor people is more vulnerable to infectious diseases, whereas a infectious diseases does not always imply poverty. Hence, the main difference between exceptional models and exceptional relationships relies on the fact that the former describes exceptional subgroups comprising a conjunction of items, whereas exceptional relationships form rules that describe strong implications between items within each exceptional subgroup. Additionally, it should be noted that the computational cost of mining exceptional relationships is higher than the one of extracting exceptional models since it is required the mining of both exceptional subgroups and association rules within each exceptional subgroup.

In a formal way, the exceptional relationships mining task can be described as follows: let us consider a dataset $\mathscr{D}$ comprising a set of transactions $t \in \mathscr{D}$, and a subgroup $G \in \mathscr{G}$ that forms an exceptional model. This exceptional behaviour is quantified according to a measure ϕ and a minimum quality threshold ϵ, i.e. $\phi^{\mathscr{D}}(G) \geq \epsilon$. The mining of exceptional relationships is defined as the task of finding all the set of association rules $\mathscr{R}$ with corresponding minimal relationship description, such that each $R \in \mathscr{R}$ describes behaviour within an exceptional subgroup G.

8.3 Algorithmic Approach

Similarly to any pattern mining algorithm, the extraction of exceptional relationships can be approached in many different ways. Nevertheless, depending on the kind of patterns to be discovered, the mining process can be challenging and different methodologies described in previous chapters may not be appropriate [1]. For instance, the use of continuous patterns by means of an exhaustive search approach is troublesome [22] due to the huge search space and many other methodologies have been considered in this regard [15, 18, 20]. Additionally, since the extraction of exceptional relationships in the intersection between exceptional models and association rules, any approach in this field inherits the drawbacks considered by any association rule mining algorithm. As described in previous chapters, the use of grammars [14] enables to overcome some of these drawbacks and provides really interesting advantages (restriction of the search space, adding subjective knowledge to the mining process, dealing with different kind of patterns, etc).

The first use of grammars for mining exceptional relationships was described by Luna et al. [16]. This evolutionary algorithm, known as MERG3P (Mining Exceptional Relationships with Grammar-Guided Genetic Programming), defines a grammar where solutions comprise a set of items in the antecedent of the rule, whereas the consequent is formed only by a single item. This grammar enables both discrete and continuous patterns to be obtained, and only a simple modification in the grammar provokes the extraction of any type of rule. MERG3P encodes each solution as a sentence of the language generated by the grammar G (see Fig. 8.4), considering a maximum number of derivations to avoid extremely large solutions. Thus, considering the grammar defined in this approach, the following language is obtained $L(G) = \{ (AND\ Condition)^n\ Condition \rightarrow Consequent : n \geq 0\}$. To obtain individuals, a set of derivation steps is carried out by applying the production rules declared in the set P and starting from the start symbol *Rule*. The way in which solutions are encoded and their representation were widely described in Chap. 5.

Analysing the definition of exceptional relationships provided in the previous section, it was described that any exceptional relationship is defined within a subgroup fitted to a specific set of target features. In this regard, each solution provided by MERG3P includes some features not included in the rule, which describe the target features. Thus, MERG3P encode solutions comprising either a

$G = (S_N, S_T, P, S)$ with:

    S = Rule
    S_N = {Rule, Conditions, Consequent, Condition, Condition_Nominal,
            Condition_Numerical }
    S_T = {'AND', 'Attribute', 'Consequent_value', '=', 'IN','Min_value','Max_value'
            'Consequent_attribute' }
    P = {Rule = Conditions, Consequent ;
            Conditions = 'AND', Conditions, Condition | Condition ;
            Condition = Condition_Nominal | Condition_Numerical ;
            Condition_Nominal = 'Attribute', '=', 'value' ;
            Condition_Numerical = 'Attribute', 'IN', 'Min_value', 'Max_value' ;
            Consequent = Condition_Nominal | Condition_Numerical ; }

Fig. 8.4 Context-free grammar to represent exceptional association rules expressed in extended BNF notation

rule represented by means of the context free grammar (see Fig. 8.4), and a string of bits where the i-th bit represents the i-th item within the dataset. In this string of bits, a value 1 in the i-th bit means that the i-th item is used as a target feature, where a value 0 means that it is not considered. Back to the previously described rule *poverty* → *infectious_disease* defined on the target features *age* and *mortality_rate*, the string of bits is represented as 1100 if the order of the attributes is as follows: *age, mortality_rate, poverty, infectious_disease*.

In a further analysis, let us study the search space for this problem. This search space might be huge since the goal is not only the discovery of association rules but also a set of target features where the rule is exceptional. In many situations, the number of both target features and rules associated to that features is extremely high and the problem can be considered as computationally intractable. In this regard, the use of an evolutionary algorithm like MERG3P (see Fig. 8.5) is highly recommended since exhaustive search approaches may suffer from both computational and memory complexity.

MERG3P starts by generating a set of solutions by means of the context-free grammar illustrated in Fig. 8.4. Once this set is obtained, the evolutionary process starts and a series of processes is carried out over generations (see Fig. 8.5, lines 10–27). In a first process, MERG3P uses a tournament selector [6] to randomly select a subset of solutions from the set *population* (see Fig. 8.5, line 11). From this set of parents, new solutions are obtained through the application of two genetic operators. The first genetic operator creates new solutions by randomly generating new conditions from the set *parents*. In this regard, this specific genetic operator randomly chooses a condition from the set of conditions of a given parent, creating a completely new condition that replaces the old-one (see Fig. 8.5, lines 12–17). Then, the second genetic operator first groups the solutions according to their target features, and those defined on the same set of target features are kept in the same group. Then, it takes the worse solution from each group and modifies the target features (see Fig. 8.5, lines 19–24).

Require: *population_size, maxGenerations* {Number of individuals and and generations}
Ensure: *population*
1: *population* ← ∅
2: *parents* ← ∅
3: *offspring* ← ∅
4: *number_generations* ← 0
5: *population* ← generate(*population_size*)
6: **for** ∀*member* ∈ *population* **do**
7:     evaluate(*member*)
8: **end for**
9: **while** *number_generations* < *maxGenerations* **do**
10:     *parents* ← select individuals from *population* by using a tournament selector
11:     **for** ∀*member* ∈ *parents* **do**
12:         *condition* ← select a random condition from *member*
13:         *newCondition* ← create a random condition
14:         *member* ← replace *condition* by *newCondition*
15:         *offspring* ← *offspring* ∪ *member*
16:     **end for**
17:     *aux_population* ← *offspring* ∪ *popluation*
18:     **for** ∀*member* ∈ *aux_population* **do**
19:         **if** *member* is the worse solution in its group of target features **then**
20:             select new target features for *member*
21:             evaluate(*member*)
22:         **end if**
23:     **end for**
24:     *population* ← select the best *population_size* solutions from *aux_population*
25:     *number_generations* + +
26: **end while**
27: return *population*

Fig. 8.5 Pseudo-code of the MERG3P algorithm

The evaluation process is another major issue in MERG3P, since it is required to quantify the exceptionalness of each rule. In this sense, the algorithm includes a evaluation procedure based on the *Pearson*'s correlation coefficient ρ (see Eq. (8.1)), determining the exceptionalness of each solution with respect to its target features. The *Pearson*'s coefficient measures the strength of linear dependence between two variables, and it is defined as the division between the covariance of the two variables and the product of their standard deviations. This coefficient ranges from 1 to −1, indicating a positive or negative correlation between the variables, respectively. A value of 0 indicates no linear correlation between the variables.

$$\rho_{X,Y} = \frac{COV(X, Y)}{\sigma_X \sigma_Y} \tag{8.1}$$

The proposed fitness function is based on three equations: *Pearson*'s correlation coefficient, support and confidence. Support and Confidence quality measures are used to determine whether a rule should be analized, since some quality thresholds are predefined by the user. If a certain association rule does not satisfy the some quality thresholds for support and confidence metrics, then a 0 fitness function value is assigned to this solution. On the contrary, the fitness value is determined by the absolute difference between the correlation coefficient of the rule for some target features $\rho_{Targets}(Rule)$ and the correlation coefficient of the complement of the rule for the same target features $\rho_{Targets}(\overline{Rule})$. Therefore, starting from an association rule R, the fitness function F used in this algorithm is defined in Eq. (8.2).

$$
F(R) = \begin{cases} |\rho_{Targets}(R) - \rho_{Targets}(\overline{R})| & \begin{aligned} &\text{if } Support_{min} \leq Support(R) \ \wedge \\ &Support(R) \leq Support_{max} \ \wedge \\ &Confidence(R) \geq Confidence_{min} \end{aligned} \\ \\ 0 & \text{otherwise} \end{cases} \tag{8.2}
$$

The final step of each generation of the evolutionary process is update the set *population*. In this regard, n individuals having a higher fitness function value are taken to form the new population. Noted that this final procedure is implemented to avoid an uncontrolled increase of the population size. In order to avoid repeated individuals, if two individuals are the same (considering both the syntax-tree and the context bit-set), then only one is considered. Only and only if the population of size n is not completed, repeated individuals could be used.

This set of previously described procedures is carried out over the generations. Once the algorithm reaches the maximum number of generations previously established by the data-miner, it returns the population set, comprising the best solutions discovered along the evolutionary process.

Finally, it is interesting to analyse the efficiency of the proposed model by means of a computational complexity analysis. In this sense, each of the main procedures of MERG3P are analysed separately: individual generator, evaluator, parent selector and genetic operator. Firstly, the computational complexity of the generator procedure depends on the maximum derivation tree ($\mathscr{X}_{max\_der}$) and the number of individuals to be created ($\mathscr{X}_{num\_ind}$). In this regard, both variables determine the complexity of this procedure, represented as $\mathscr{O}(\mathscr{X}_{max\_der} \times \mathscr{X}_{num\_ind})$, its running time increases linearly with the maximum tree size and the number of individuals. As for the evaluator procedure, it depends on the number of individuals ($\mathscr{X}_{num\_ind}$), transactions ($\mathscr{X}_{num\_trans}$) and items ($\mathscr{X}_{num\_items}$). Mathematically, this value is defined as $\mathscr{O}(\mathscr{X}_{num\_trans} \times \mathscr{X}_{num\_items} \times \mathscr{X}_{num\_ind})$. Additionally, the complexity order of the parent selector procedure is constant since it only requires a random value, i.e., $\mathscr{O}(1)$. Finally, the computational complexity of the genetic operator procedure depends on both the derivation tree size ($\mathscr{X}_{max\_der}$) and number of individuals ($\mathscr{X}_{num\_ind}$). In consequence, its complexity order is defined as $\mathscr{O}(\mathscr{X}_{max\_der} \times \mathscr{X}_{num\_ind})$. Analysing the computing requirements for each procedure, it is stated that both $\mathscr{X}_{max\_der}$ and $\mathscr{X}_{num\_ind}$ are previously fixed, so they are considered as constants and their complexity order is $\mathscr{O}(1)$. Additionally, all the procedures are repeated as many times as the predefined number of generations, which is also a constant value predefined. Therefore, bearing in mind all these issues, the resultant computational complexity of MERG3P is stated as $\mathscr{O}(\mathscr{X}_{num\_trans} \times \mathscr{X}_{num\_items})$. Thus, the complexity of the MERG3P approach is linear with regard to the number of transactions and the number of items.

8.4 Successful Applications

The aim of this section is to present a really interesting real application field where the extraction of exceptional relationships is of high interest. In this regard, this section deals with the discovery and analysis of hidden exceptional relations between quality measures in the pattern mining field. This analysis is of high interest to understand the behaviour of the metrics and to be able to choose the correct measures to be optimized. Thereby, a dataset comprising all the available values for each measure is used so that any exceptional relation between measures could be discovered. The benchmark dataset used in this study comprises any available value for a set of quality measures: Support, Support of the antecedent, Support of the consequent, Confidence, Lift, Conviction, Leverage, Certainty Factor and Cosine.

When analysing the aforementioned quality measures, a series of exceptional association rules were discovered. One of the most interesting exceptional rules—based on its fitness value, which is equal to 0.9631—is defined for the target features support of the consequent and conviction (see Fig. 8.6). The rule obtained is defined as **IF** *confidence* IN [0.2, 0.4] **THEN** *cosine* IN [0.1, 0.8], which is satisfied in 12.75 % of the transactions and it is a high accurate rule, since its confidence is almost maximum, i.e. 97.01 %. This exceptional association rule describe that, despite the fact that conviction and support of the consequent seem to be independent, the correlation between them is negative if confidence takes a value in the range [0.2, 0.4] and cosine is defined in the range [0.1, 0.8].

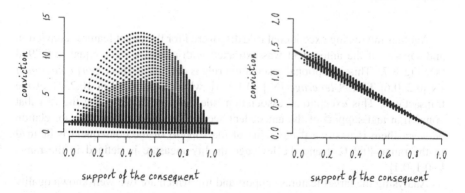

Fig. 8.6 Example of the scatter plot and regression line for the rule **IF** *confidence* IN [0.2, 0.4] **THEN** *cosine* IN [0.1, 0.8] defined on the target features conviction and support of the consequent (*right*) and its complement (*left*)

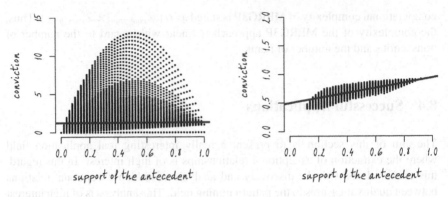

Fig. 8.7 Example of the scatter plot and regression line for the rule **IF** *support consequent* IN [0.2, 0.6] **THEN** *leverage* IN [−0.1, 0.1] defined on the target features conviction and support of the antecedent (*right*) and its complement (*left*).

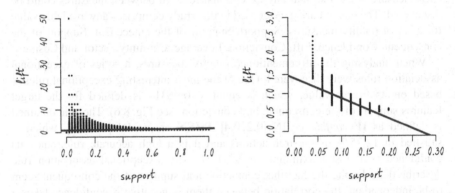

Fig. 8.8 Example of the scatter plot and regression line for the rule **IF** *support antecedent* IN [0.4, 0.7] **THEN** *cosine* IN [0.2, 0.3] defined on the target features support and lift (*right*) and its complement (*left*)

Another interesting exceptional rule discovered for the target features conviction and support of the antecedent was extracted with a fitness value equal to 0.7928 (see Fig. 8.7). This exceptional association rule is defined as **IF** *support consequent* IN [0.2, 0.6] **THEN** *leverage* IN [−0.1, 0.1], which is satisfied in 10.23 % of the transactions. This exceptional association rule describe that, despite the fact that conviction and support of the antecedent seem to be independent, the correlation between them is positive if the value of the support of the consequent is defined in the range [0.2, 0.6] and the leverage quality measure is defined in the range [−0.1, 0.1].

Analysing the target features support and lift, which are two well-known quality measures in the pattern mining field, we obtain that they are almost positively correlated. Nevertheless, this correlation is negative when support of the antecedent and cosine measures come to play (see Fig. 8.8). Thus, an exceptional association rule is

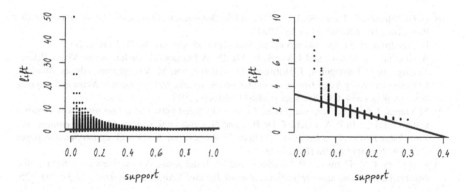

Fig. 8.9 Example of the scatter plot and regression line for the rule **IF** *leverage* IN [0.0, 0.2] **THEN** *cosine* IN [0.1, 0.6] defined on the target features support and lift (*right*) and its complement (*left*)

discovered, which is defined as **IF** *support antecedent* IN [0.4, 0.7] **THEN** *cosine* IN [0.2, 0.3], which means that the distribution of the transactions is completely different on the features support and lift when the pattern {*support antecedent* IN [0.4, 0.7], *cosine* IN [0.2, 0.3]} is considered.

Finally, on the same set of target features, i.e. support and lift quality measures, the algorithm MERG3P discovers another interesting exceptional association rule or exceptional behaviour in data. When the distribution of the transactions for target features is analysed, a regression coefficient value of 0.1431 is obtained, meaning that both measures are positive correlated but the slope is slow. On the contrary, considering the rule **IF** *leverage* IN [0.0, 0.2] **THEN** *cosine* IN [0.1, 0.6] on the same target features (see Fig. 8.9), the value obtained for the regression coefficient is -0.6669, determining a negative correlation for the target features support and lift.

References

1. C. C. Aggarwal and J. Han. *Frequent Pattern Mining*. Springer International Publishing, 2014.
2. R. Agrawal, T. Imielinski, and A. N. Swami. Mining association rules between sets of items in large databases. In *Proceedings of the 1993 ACM SIGMOD International Conference on Management of Data*, SIGMOD Conference '93, pages 207–216, Washington, DC, USA, 1993.
3. M. Atzmueller. Subgroup Discovery - Advanced Review. *WIREs: Data Mining and Knowledge Discovery*, 5:35–49, 2015.
4. W. Duivesteijn, A. J. Knobbe, A. Feelders, and M. van Leeuwen. Subgroup discovery meets bayesian networks – an exceptional model mining approach. In *Proceedings of the 2010 IEEE International Conference on Data Mining*, ICDM 2010, pages 158–167, Sydney, Australia, December 2010. IEEE Computer Society.
5. W. Duivesteijn, A. J. Feelders, and A. Knobbe. Exceptional model mining. *Data Mining and Knowledge Discovery*, 30(1):47–98, 2015.

6. D. Dumitrescu, B. Lazzerini, L. C. Jain, and A. Dumitrescu. *Evolutionary Computation.* CRC Press, Inc., Boca Raton, FL, USA, 2000.

7. D. Freedman, R. Pisani, and R. Purves. *Statistics (4th edition).* W. W. Norton, 2007.

8. B. Goethals, S. Moens, and J. Vreeken. MIME: A Framework for Interactive Visual Pattern Mining. In D. Gunopulos, T. Hofmann, D. Malerba, and M. Vazirgiannis, editors, *Machine Learning and Knowledge Discovery in Databases,* volume 6913 of *Lecture Notes in Computer Science,* pages 634–637. Springer Berlin Heidelberg, 2011.

9. M. Gupta, J. Gao, Y. Sun, and J. Han. Community trend outlier detection using soft temporal pattern mining. In P. A. Flach, T. De Bie, and N. Cristianini, editors, *Machine Learning and Knowledge Discovery in Databases,* volume 7524 of *Lecture Notes in Computer Science,* pages 692–708. Springer Berlin Heidelberg, 2012.

10. F. Herrera, C. J. Carmona, P. González, and M. J. del Jesus. An overview on subgroup discovery: Foundations and applications. *Knowledge and Information Systems,* 29(3): 495–525, 2011.

11. M. Leeuwen and A. Knobbe. Diverse subgroup set discovery. *Data Mining Knowledge Discovery,* 25(2):208–242, 2012.

12. D. Leman, A. Feelders, and A. J. Knobbe. Exceptional model mining. In *Proceedings of the European Conference in Machine Learning and Knowledge Discovery in Databases,* volume 5212 of *ECML/PKDD 2008,* pages 1–16, Antwerp, Belgium, 2008. Springer.

13. J. M. Luna, A. Ramirez, J. R. Romero, and S. Ventura. An intruder detection approach based on infrequent rating pattern mining. In *Proceedings of the 10th International Conference on Intelligent Systems Design and Applications, ISDA 2010,* ISDA 2010, pages 682–688, 2010.

14. J. M. Luna, J. R. Romero, and S. Ventura. Design and behavior study of a grammar-guided genetic programming algorithm for mining association rules. *Knowledge and Information Systems,* 32(1):53–76, 2012.

15. J. M. Luna, J. R. Romero, C. Romero, and S. Ventura. Reducing gaps in quantitative association rules: a genetic programming free-parameter algorithm. *Integrated Computer Aided Engineering,* 21(4):321–337, 2014.

16. J. M. Luna, M. Pechenizkiy, and S. Ventura. Mining exceptional relationships with grammar-guided genetic programming. *Knowledge and Information Systems,* pages 1–24, In press,2015.

17. Y. Z. Ma. Simpson's paradox in GDP and per capita GDP growths. *Empirical Economics,* 49(4):1301–1315, 2015.

18. D. Martín, A. Rosete, J. Alcalá, and F. Herrera. A new multiobjective evolutionary algorithm for mining a reduced set of interesting positive and negative quantitative association rules. *IEEE Transactions on Evolutionary Computation,* 18(1):54–69, 2014.

19. R. McKay, N. Hoai, P. Whigham, Y. Shan, and M. O'Neill. Grammar-based Genetic Programming: a Survey. *Genetic Programming and Evolvable Machines,* 11:365–396, 2010.

20. R. Srikant and R. Agrawal. Mining Quantitative Association Rules in Large Relational Tables. In *Proceedings of the 1996 ACM SIGMOD International Conference on Management of Data,* SIGMOD'96, Montreal, Quebec, Canada, 1996.

21. M. van Leeuwen. Maximal exceptions with minimal descriptions. *Data Mining and Knowledge Discovery,* 21(2):259–276, 2010.

22. C. Zhang and S. Zhang. *Association rule mining: models and algorithms.* Springer Berlin / Heidelberg, 2002.

Chapter 9
Scalability in Pattern Mining

Abstract The pattern mining task is the keystone of data analysis, describing and representing any type of homogeneity and regularity in data. Abundant research studies have been dedicated to this task, providing overwhelming improvements in both efficiency and scalability. Nevertheless, the growing interest in data collection is giving rise to extremely large datasets that hinder the mining process. Thus, it is essential to provide solutions to the challenges derived from the processing of such high dimensional datasets in an efficient way. This chapter aims to describe different ways of speeding up the pattern mining process, presenting some traditional methods for handling very large data collections, and new trends in the mining of patterns in Big Data.

9.1 Introduction

Nowadays, the increasing amount of data generated in different fields have provoked an increasing interest in extracting, analysing and taking advantage of the knowledge hidden in such datasets. As a matter of example, Internet has become an indispensable tool for business, generating information about users and enabling the discovery of new trends and shopping habits. In this sense, the increasing interest in data storage has given rise to increasingly large datasets [35], which may exceed the capacity of traditional techniques with respect to memory consumption and performance in a reasonable quantum of time [15]. In many fields, the extraction and analysis of useful and unknown information from raw data is a crucial activity and, sometimes, this procedure cannot take more than few seconds. Thus, the data processing and data analysis of huge datasets to extract unknown knowledge from raw data in an efficient way is more and more interesting [42], and speeding up the data accesses is cornerstone. Many studies have been focused on different data mining techniques [43] considering the scalability and performance issues.

The task of mining frequent patterns is one of the most well-known and intensively researched problems in data mining [22]. In this field, numerous algorithms have been designed and developed to solve problems related to computational and memory requirements. For a better understanding of the pattern mining problem, let us consider a brute-force approach for mining any existing pattern in a dataset comprising N transactions and k single items. In such a dataset, the total number

© Springer International Publishing Switzerland 2016
S. Ventura, J.M. Luna, *Pattern Mining with Evolutionary Algorithms*,
DOI 10.1007/978-3-319-33858-3_9

of patterns that might be generated from k single items is equal to $M = 2^k - 1$, which exponentially increases with the number k of single items. Thus, the number of both transactions and single items in a dataset have a negative impact in the computational complexity since $M \times N$ comparisons are required to determine the frequency of occurrence of each pattern $M_i \in M$. Existing efficient algorithms in the pattern mining field follow some frequent item-set generation strategies [13], which are based on reducing either the number M of candidate sets or the number N of transactions.

Let us focus now on the problem of mining association rules from the set of M patterns. Considering a dataset comprising k singletons, it is possible to obtain a total of $3^k - 2^{k+1} + 1$ different association rules. There are $\binom{k}{2}$ itemsets of length 2, and each of these itemsets can produce $2^2 - 2 = 2$ association rules; there are also $\binom{k}{3}$ itemsets of length 3 and each of them can produce $2^3 - 2 = 6$ association rules, and so on. This process is repeated until $\binom{k}{k}$, which produces $2^k - 2$ different association rules. The process of determining patterns and associations between patterns requires a high computational cost and huge amount of main memory, specially when the datasets used in the mining process become bigger and bigger. For such datasets, the importance of pruning the search space is primordial, given rise to the paradigm of constraint-based mining [3]. The mining process is carried out by means of a set of constraints that capture application semantics. It helps users' exploration and control, and the paradigm allows to confine the search space of patterns to those of interest to the users achieving superior performance.

For the sake of reducing the computational time and the memory requirements, different metaheuristics have been proposed to solve the pattern mining problem. Many researchers have achieved some promising results when using evolutionary algorithms [29, 31, 37], which extract association rules without requiring a prior step for mining frequent items as most exhaustive search algorithms do [47]. Some evolutionary algorithms in the association rule mining problem were developed by means of genetic algorithms (GAs). The use of GAs is considered by some authors as a way of optimizing and looking for patterns of interest [36, 45], so thanks to this type of algorithms, the pattern mining problem can be considered as a combinatorial optimization problem where solutions are represented as strings of genes and each gene describes the way in which a specific item is analysed [12]. Grammar-guided genetic programming (G3P) [38] has also been applied to the extraction of association rules [27], where the use of a grammar allows for defining syntax constraints, restricting the search space, and obtaining expressive solutions in different attribute domains. The use of grammars can be considered as a useful way for introducing subjective knowledge to the pattern mining process, enabling patterns of interest to be obtained by users with different goals. Grammars are highly related to the background knowledge of the user [20], and they allow to define solutions that have flexible and expressive structures that ease the interpretability of the extracted knowledge.

Even when the use of different metaheuristics has overcome most of the existing problems in the pattern mining field, the growing interest in data collection and the resulting set of large datasets to be analysed are hindering the mining process [39]. In this sense, parallel frequent pattern mining is emerging as an interesting research area where many hierarchical parallel environments with multi-core processors and graphic processing units (GPUs) have reduced the computational time required during the mining process. In the same way, new trends are based on the term Big Data, which refers to the challenges and advances derived from the processing of such high dimensional datasets in an efficient way. In this regard, it is becoming essential to design parallel and distributed methods capable of handling very large data collections in a reasonable time [5], new efficient pattern mining algorithms based on the MapReduce framework [11], or even new data representations that improve the data accesses [30].

9.2 Traditional Methods for Speeding Up the Mining Process

The increasing amount of data gathered in different fields have provoked extremely large datasets to be analysed. These huge datasets hamper the mining process and slowing down the performance of proposed algorithms due to the large number of both single items and transactions. In this section, the importance of evolutionary computation in scalability issues is described. Additionally, new parallel algorithms for mining association rules are described, which speed up the evaluation procedure and reduce the time required in the mining process. Finally, this section outlines some new data representations that improve the data accesses so the time required for the mining process is highly reduced.

9.2.1 The Role of Evolutionary Computation in Scalability Issues

The extraction of patterns of interest and associations between them can be treated as a combinatorial optimization problem [47]. Considering a brute-force approach for mining any existing pattern in a dataset comprising k single items, a total of $2^k - 1$ patterns or combinations of items can be obtained, and a set containing $3^k - 2^{k+1} + 1$ different association rules can be discovered. Therefore, the search space in pattern mining exponentially increases with the number k of single items. All of this lead us to the conclusion that the pattern mining problem is a really complex task, and this complexity depends on the number of both single items and transactions. The increasing number of items provokes an increment in both the computational time and the memory requirements, whereas the number of transactions is more related to the computational complexity since any extracted pattern needs to be evaluated for each single transaction.

The pattern mining task can be formulated as an optimization problem where the aim is to obtain a set of patterns that satisfy some specific quality measures. In this regard, the use of an evolutionary perspective for the aforementioned task has been considered for many researchers [8, 12], extracting optimal patterns based on a concrete fitness functions defined on the basis of different quality measures [8]. As described by many authors [24, 34], the use of an evolutionary perspective for the pattern mining problem in different domains enables the computational and memory requirements to be tackled, being really suitable in situations where large search spaces are required to be explored. The improvement of the performance of pattern mining algorithms is highly related to genetic operators used in evolutionary process, which enable to guide the searching process to promising regions of the search space, and to discard those areas that seem to be meaningless due to solutions around them are far to be optimal.

The flexibility to represent solutions also plays an important role in scalability issues, enabling the search space to be reduced to those patterns that satisfy the predefined constraints. Evolutionary computation is also really suitable in situations where the flexibility of the representation of the solutions is a dare. This flexibility is specially important when the search space is highly dependent on both the data domain and the type of patterns to be extracted [3, 47], including continuous, discrete, fuzzy representations, frequent, rare, positive, negative, maximal, colossal, spatio-temporal, etc.

Many researchers have focused their studies on mining patterns of interest by means of metaheuristics [29, 31, 37]. Some algorithms in the pattern mining field have been developed by means of genetic algorithms (GAs), which are able to extract association rules without requiring a prior step for mining frequent items as most exhaustive search algorithms do [47]. GAs have emerged as robust and practical optimization methods [16], and they are generally known for their powerful problem solving capabilities and their flexibility in design, which makes them suitable for a wide range of problems and data domains. GAs are considered as really efficient models, obtaining the solution, or the set of them, in a reasonable amount of time and providing a good balance between solution quality and response time.

The efficiency and the quality of the solutions of any GA might be determined as the product of two factors: the population size and the number of generations it takes to converge. The population set comprises solutions to the problem, which are generally encoded as strings of values in the pattern mining field, and the length and structure of the solutions vary depending on the information to be represented [14]. As for the convergence issue, it is determined by the selection pressure so an extremely fast convergence may be detrimental to the solution quality.

Despite the fact that different authors have described the good performance of GAs in the pattern mining field [12], the way in which their solutions are encoded hampers the restriction of the search space beforehand. Most GAs in this field look for patterns of interest in the whole search space, and some minor restrictions can be considered in terms of size of the patterns and consequent part of association rules. Existing proposals based on GAs do not consider external knowledge to be used, so the subjectivity is only considered by means of some subjective quality measures like, for instance, comprehensibility [14]. Considering the restriction of the search space as a baseline to enhance the performance of the evolutionary algorithm, the use of grammars to include syntax constraints is a really interesting point. Nevertheless, the use of grammars comes with the consequent risk of not reaching the optimal solutions due to the grammars' constraints [38], so convergence of the algorithm needs to be postponed. Grammars have been successfully applied to different approaches for mining patterns and associations of interest [28, 29], reducing the search space and introducing subjective knowledge to the pattern mining process.

Finally, it should be noted that, even when the use of evolutionary algorithms in the pattern mining field has reduced the problems dealing with computational time, memory requirements and reduction of the search space, all these problems emerge again when extremely large datasets are considered. In this sense, it is becoming essential to design parallel and distributed evolutionary algorithms capable of handling such high dimensional datasets in an efficient way and in a reasonable quantum of time [5]. Additionally, these new challenges might be faced up by means of new efficient pattern mining algorithms based on the MapReduce framework [11], or even new data representations that improve the data accesses [30].

9.2.2   Parallel Algorithms

Many algorithms have been developed for solving the pattern mining problem by considering different perspectives. First approaches were based on an exhaustive search methodology, Apriori [4], FP-Growth [23] and Eclat [46] are the most well-known algorithms in this sense. With the increasing interest in mining patterns and association between them in different application fields, there is an increment in both the size of the datasets and the type of patterns to be extracted. Many different pattern mining algorithms based on evolutionary computation [32, 40] have been proposed to provide solutions to the increasing and varied number of datasets. However, when dealing with extremely large amount of data, the complexity time might turns unmanageable.

Parallel frequent pattern mining is emerging as an interesting research area where many hierarchical parallel environments with multi-core processors and graphic processing units (GPUs) have reduced the computational time required during the mining process [17, 48]. The use of GPUs has already been studied in machine learning [9, 21], and specifically for speeding up algorithms within the framework of evolutionary computation [25]. GPU consists of a large number of processors and recent devices operate as multiple instruction multiple data (MIMD), and also as single instruction multiple data (SIMD) architectures.

First proposals for mining frequent patterns in a parallel way were based on well-known exhaustive search algorithms. Two of the first proposals [44] in this sense were designed as two parallel versions of Apriori called PBI

(Pure Bitmap Implementation) and TBI (Trie Bitmap Implementation), respectively. These GPU-based implementations of Apriori employ a bitmap data structure to encode the dataset on the GPU and utilize the GPU's SIMD (Single Instruction, Multiple Data) parallelism for support counting.

Adil et al. [2] proposed a new algorithm for mining association rules based on a GPU architecture that works into two steps. First, it carries out a generation of frequent itemsets where each thread block computes the support of a set of itemsets. Second, the obtained set of frequent itemsets are sent back to the CPU in order to generate association rules. One of the main drawbacks of this algorithm is the cost of the CPU/GPU communications.

Another parallel algorithm for mining frequent patterns was proposed by Cui et al. [10]. This algorithm uses CUDA (Computer Unified Device Architecture), a parallel computing architecture developed by NVIDIA that allows programmers to take advantage of the computing capacity of NVIDIA GPUs in a general purpose manner. The CUDA programming model executes kernels as batches of parallel threads in a SIMD programming style. These kernels comprise thousands to millions of lightweight GPU threads per each kernel invocation. The proposed algorithm, which is based on Apriori, first divides the dataset among different threads. Then, k-candidate itemsets are generated and allocated on global memory. Each thread handles one candidate using only the portion of the dataset assigned to its block. In each iteration, a synchronization between blocks is done in order to compute the global support of each candidate.

To date, there is only one work that introduces metaheuristics for association rule mining on GPU [6]. When analysing the process of computing quality measures for different solutions (association rules) in an evolutionary algorithm, it is discovered that there are no internal dependencies between solutions so different rules might be evaluated concurrently. Furthermore, for each association rule, the computing process between the antecedent and consequent of a rule is also independent, so the support of the antecedent and consequent parts of each rule might also be computed in a concurrent way. All of this lead us to the conclusion that the evaluation process of any association rule on any quality measure provides a high degree of parallelism.

The work proposed in [6] is based on the fact that the use of evolutionary algorithms in the field of association rule mining requires a hard process for evaluating each rule on datasets comprising a large number of transactions. Note that each association rule is evaluated for each transaction to compute the frequency of occurrence. Thus, the higher the number of transactions to be checked, the higher the computational time for the evaluation process. In fact, different studies carried out by authors have demonstrated that more than 95 % of the time is consumed in evaluating the solutions. Therefore, reducing the time required for the evaluation process enables a high reduction in the mining process, and GPUs are efficient and high-performance platforms to this aim.

Cano et al. [6] proposed a GPU model designed to evaluate association rules on approaches based on metaheuristics. In this regard, the proposed GPU model comprises two kernels for computations of the coverage and reduction, and a final kernel to compute the fitness value. Figure 9.1 shows the way in which different

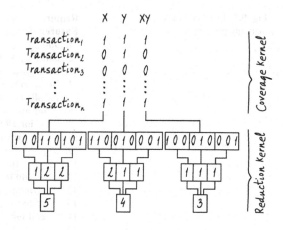

Fig. 9.1 Example of evaluation model including coverage kernel and reduction kernel

transactions are evaluated for the antecedent X, consequent Y, and both XY. First, the rule, antecedent and consequent, are copied to the GPU memory. Specifically, they are copied to the constant memory, which provides broadcast to the GPU threads. Second, the coverage kernel is executed for the antecedent X and the consequent Y, using the transactions of the dataset. A value one is obtained if the transaction is satisfied and zero otherwise. Once the bitsets for X and Y are obtained, the bitset of XY will be determined by the intersection of X and Y. Third, the reduction kernel is executed and the global supports for X, Y and XY are calculated. To this end, the bitsets are splitted into portions and partial results are calculated. The process continues till the global support is obtained.

A major feature of this proposal is that kernel executions might be overlapped so the coverage kernel may overlap its execution when concurrently evaluating the antecedent X and the consequent Y. Note that both X and Y operate with different data and they have no dependencies, as previous described. This feature is also considered by the reduction kernel, since X, Y and XY can be reduced concurrently since they are independent.

9.2.3 New Data Structures

As previously described, the increasing amount of electronically available data have given rise to extremely large datasets that exceed the capacity of different machine learning algorithms in obtaining solutions in a reasonable quantum of time [15]. First proposals for speeding up the pattern mining problem [19] were focused on models that either reduced the search space and used a highly efficient representation of the frequent itemsets. However, it is also possible to go one step further and to propose new suitable data structures that enable to handle and fast compute extremely large datasets without any adaptation of the algorithms'

Fig. 9.2 Inverted index
mapping procedure

Require: *sorted_data*
Ensure: $data_{index\_based}$
1: $data_{index\_based} \leftarrow \emptyset$
2: $index_{value} \leftarrow \emptyset$
3: $index_{attribute} \leftarrow \emptyset$
4: $list\_attributes \leftarrow getAttributes(sorted\_data)$
5: **for all** *attribute* in *list_attributes* **do**
6:     **for all** *value* in *attribute* **do**
7:         **for all** *transaction* in *sorted_data* **do**
8:             **if** *value* satisfies the *i*-th *transaction* **then**
9:                 $index_{value} \leftarrow index_{value} \cup i$
10:            **end if**
11:        **end for**
12:        $index_{attribute} \leftarrow index_{attribute} \cup index_{value}$
13:        $index_{value} \leftarrow \emptyset$
14:    **end for**
15:    $data_{index\_based} \leftarrow data_{index\_based} \cup index_{attribute}$
16:    $index_{attribute} \leftarrow \emptyset$
17: **end for**
18: **return** $data_{index\_based}$

methodology. These new data structures are able to simplify and reorganize data items in order to reduce the data size and provide a faster access to the stored information.

Many of these data structures are based on the fact that the number of different items in a datasets is not so high. As a matter of example, let us consider that a specific numerical item is different from transaction to transaction. Then, the number of distinct values observed in a specific item is in the range of $[1, N]$, considering N as the number of transactions. Considering all of this, it is possible to reduce the memory storage by sorting data in a special way by using a shuffling strategy based on the hamming distance (HD). Some authors have employed inverted index mapping [26], run length encoding (RLE) compression [1], etc, to re-organize items within data allowing the whole dataset to be fit into main memory, and achieving faster data accesses.

Recently, Luna et al. [30] proposed a highly efficient data structure to be used on any pattern mining algorithms and, specially, evolutionary algorithms for mining association rules. The proposed data representation efficiently processes multiple heterogeneous datasets comprising attributes of any domain, either discrete and continuous, and ensuring the data validity. The aim of this new data representation is to reorganize data items by means of a shuffling strategy based on HD as a similarity metric. Transactions are grouped in order to minimize the number of changes in the values of the items from transaction to transaction.

Once the shuffling strategy is carried out, an inverted index mapping procedure is run (see Fig. 9.2). This inverted index mapping procedure aims at defining a new data structure based on indices ($data_{index\_based}$) and considering the sorted data previously obtained from the shuffling strategy. It assigns a key-value pair to each list of attributes, providing a new data structure where the attribute values have

Fig. 9.3 Run length
encoding compression

Require: $data_{index\_based}$
Ensure: $data_{structure}$
1: $data_{structure} \leftarrow \emptyset$
2: $index_{value} \leftarrow \emptyset$
3: $index_{attribute} \leftarrow \emptyset$
4: $list\_attributes \leftarrow getAttributes(data_{index\_based})$
5: **for all** $attribute$ in $list\_attributes$ **do**
6:     **for all** $value$ in $attribute$ **do**
7:         **if** $value$ comprises consecutive indices **then**
8:             $index_{value} \leftarrow$ getCompressedIndices($value$)
9:         **else**
10:            $index_{value} \leftarrow$ getIndices($value$)
11:        **end if**
12:        $index_{attribute} \leftarrow index_{attribute} \cup index_{value}$
13:        $index_{value} \leftarrow \emptyset$
14:    **end for**
15:    $data_{structure} \leftarrow data_{structure} \cup index_{attribute}$
16:    $index_{attribute} \leftarrow \emptyset$
17: **end for**
18: **return** $data_{structure}$

pointers to any transaction index covered. The procedure makes an analysis of the whole set of values for each attribute in the dataset (see lines 5–17, Fig. 9.2). If the i-th transaction is covered by the attribute value, then its index i is added to the set of index for the attribute value (see lines 8–10, Fig. 9.2).

Finally, once the data structure is obtained by the inverted index mapping procedure, a RLE compression of such a structure is carried out (see Fig. 9.3). RLE is a form of data compression in which sequences of data having the same values occur in a consecutive way. This compression procedure simplifies the previous data structure, enabling lower memory requirements and speeding up the data accesses. Beginning with the dataset previously created by the inverted index mapping procedure, this RLE compression procedure analyses each attribute value and its indices (see lines 5–17, Fig. 9.3). If a specific attribute value includes consecutive indices, then a compression is applied defining a starting index and the total number of consecutive transactions containing the same feature value {*index, displacement*}.

9.3 New Trends in Pattern Mining: Scalability Issues

The applicability of the pattern mining task to really large datasets is not an easy problem and numerous challenges are required to be solved, most of them are related to scalability and main memory. As previously introduced, the pattern mining task is one of the most well-known and intensively researched problems in data mining [22], which require a high computational time due to the large search space produced by typical datasets. Noted that the total number of patterns that might be generated from k single items in a dataset is equal to $2^k - 1$, so the search space increases exponentially with the number k of single items. In this scenario, the extraction of patterns of interest in extremely large datasets by using a single machine may not complete the mining process, and the pattern mining task should be adapted to emerging technologies [35].

One of these emerging technologies is MapReduce [11], a novel programming model that has become very popular for intensive computing. This programming model, in conjunction with its distributed file system [18] introduced by Google, offers a simple and robust method of writing distributed programs for processing large data sets. MapReduce allows to write parallel programs in a simple way and to tackle large datasets that require data-intensive computing like in the pattern mining problem. MapReduce programs are composed of two main phases defined by the programmer: map and reduce. The map phase processes a sub-set of input data and produces a set of $\langle k, v \rangle$ pairs. All the values associated with the same key k are merged and the reduce phase is responsible for working on these sets to produce the final values. It should be noted that all the map and reduce operations are run in parallel, and any inputs/outputs are stored in a distributed file system that is accessible from any computer defined to be used by MapReduce.

For a better understanding of the MapReduce framework, let us consider the sample market basket dataset (see Table 9.1) described in Chap. 1. Each transaction corresponds to a different customer that contains a set of items as products purchased by the customer. Additionally, let us also calculate the frequency of occurrence for each single item in the aforementioned dataset by using a MapReduce model (see Fig. 9.4). In this regard, the dataset is divided into subsets of data, and each mapper receives one of these subsets. The map phase is responsible for iterating over each item producing the pair $\langle k, v \rangle$, where k states for the name of the item in the dataset, and v means the number of times that k appear in the subset. Finally, the reduce phase calculates the sum of v for each item k, obtaining the number appearances of each k.

First proposals for mining patterns of interest on the MapReduce framework were based on an exhaustive search methodology. Moens et al. [39] proposed the Dist-Eclat algorithm, an approach that works into three different steps and each of these steps are distributed among multiple mappers. In the first step, the dataset is divided into equally-sized blocks and distributed to the mappers. Here, each mapper extracts the frequent singletons, i.e. items of size one, from its block. In a second step, the algorithm distributes across the mappers all the frequent singletons extracted, and

Table 9.1 An example of market basket dataset	Customer	Items
	ID_1	{Banana, Bread, Butter, Orange}
	ID_2	{Banana, Bread, Milk}
	ID_3	{Banana, Butter, Milk}
	ID_4	{Bread, Butter, Milk, Orange}
	ID_5	{Banana, Butter, Milk, Orange}

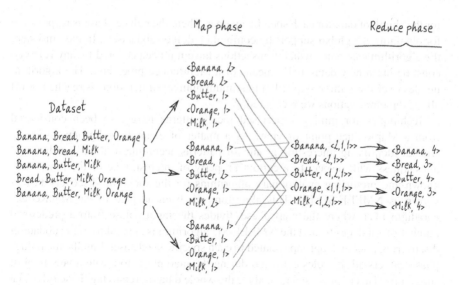

Fig. 9.4 MapReduce framework applied to the sample dataset shown in Table 9.1

the aim is to find frequent itemsets of size k formed by the singletons. All this information is expressed as a prefix tree so frequent patterns can be extracted in a final step.

One of the main drawbacks of Dist-Eclat is that the mining k-sized item-sets might be unfeasible. In this sense, Moens et al. [39] also proposed a different model, known as BigFIM, based on breadth-first. Each mapper of BigFIM receives a subset of the whole dataset and all the patterns for which we want to know the support are returned. The reducer combines all local frequencies and report only the global frequent patterns. These frequent patterns are distributed to all mappers to act as candidates for the next step of the breadth-first search. It should be noted that BigFIM is an iterative algorithms that needs to be repeated k times to obtain the patterns of size k.

Recently, some authors are describing evolutionary algorithms based on the MapReduce paradigm for mining patterns of interest. MOPNAR-BigData [33] is one of theses algorithms, which follows an iterative methodology for mining negative, positive and quantitative association rules (rules including continuous patterns). This algorithm is based on the MOPNAR [31] multi-objective algorithm, which obtained really interesting results in the pattern mining field. MOPNAR-BigData works into three different steps based on the MapReduce framework. In the first step, the input dataset is divided into a number of independent subsets and each computer node run the MOPNAR algorithm over a specific subdataset, obtaining a set of association rules of interest. All the sets of rules obtained for each subdataset are joined into a resulting set of rules. In a second step, the resulting set of rules is analysed and each rule is evaluated for the whole dataset. In this regard, a map phase is used to evaluate the support of the rule, antecedent and consequent of each rule for the chunk of data for each specific mapper. Then, the reduce phase is responsible for obtaining the global support by combining all the subdatasets. In the third step, the algorithm calculates which transactions have not been covered by any rule yet, constructing a new datasets by means of a MapReduce procedure. The algorithm iterates over these three steps till a maximum number of iterations is reached or till all the data transactions were covered.

Techniques for mining supervised local patterns have also been considered from a MapReduce point of view. As a matter of example, subgroup discovery aims at identifying interesting groups of patterns according to their distributional unusualness with respect to a certain property of interest. One of the most well-known and promising evolutionary algorithms for the subgroup discovery task is known as NMEEF-SD [7]. This algorithm has been adapted to the MapReduce paradigm [41], where the map phase divides the input dataset into a predefined number of subdatasets, and the NMEEF-SD algorithm is applied to each subdataset discovering a set of subgroups related to each specific subdataset. Finally, the reduce phase gathers all the rules extracted during the map phase to create a new pool of subgroups. This pool is used to analyse the whole dataset, removing those rules that are equal or represent the same knowledge.

References

1. D. J. Abadi, S. Madden, and M. Ferreira. Integrating compression and execution in column-oriented database systems. In *Proceedings of the ACM SIGMOD International Conference on Management of Data*, SIGMOD Conference, pages 671–682, Chicago, Illinois, USA, 2006.
2. S. H. Adil and S. Qamar. Implementation of association rule mining using CUDA. In *Proceedings of the 2009 International Conference on Emerging Technologies*, ICET 2009, pages 332–336, Islamabad, Pakistan, 2009.
3. C. C. Aggarwal and J. Han. *Frequent Pattern Mining*. Springer International Publishing, 2014.
4. R. Agrawal, T. Imielinski, and A. N. Swami. Mining association rules between sets of items in large databases. In *Proceedings of the 1993 ACM SIGMOD International Conference on Management of Data*, SIGMOD Conference '93, pages 207–216, Washington, DC, USA, 1993.
5. E. Alba and M. Tomassini. Parallelism and evolutionary algorithms. *IEEE Transactions on Evolutionary Computation*, 6(5):443–462, 2002.
6. A. Cano, J. M. Luna, and S. Ventura. High performance evaluation of evolutionary-mined association rules on gpus. *The Journal of Supercomputing*, 66(3):1438–1461, 2013.
7. C. J. Carmona, P. González, M. J. del Jesus, and F. Herrera. NMEEF-SD: Non-dominated multiobjective evolutionary algorithm for extracting fuzzy rules in subgroup discovery. *IEEE Transactions on Fuzzy Systems*, 18(5):958–970, 2010.
8. C. J. Carmona, P. González, M. J. del Jesus, and F. Herrera. Overview on evolutionary subgroup discovery: analysis of the suitability and potential of the search performed by evolutionary algorithms. *Wiley Interdisciplinary Reviews: Data Mining and Knowledge Discovery*, 4(2): 87–103, 2014.
9. J. M. Cecilia, A. Nisbet, M. Amos, J. M. García, and M. Ujaldón. Enhancing GPU parallelism in nature-inspired algorithms. *Journal of Supercomputing*, 63(3):773–789, 2013.
10. Q. Cui and X. Guo. Research on parallel association rules mining on GPU. In *Proceedings of the 2nd International Conference on Green Communications and Networks*, GCN 2012, pages 215–222, Gandia, Spain, 2012.

11. J. Dean and S. Ghemawat. MapReduce: simplified data processing on large clusters. *Communications of the ACM*, 51(1):107–113, 2008.
12. M. J. del Jesús, J. A. Gámez, P. González, and J. M. Puerta. On the discovery of association rules by means of evolutionary algorithms. *Wiley Interdisciplinary Reviews: Data Mining and Knowledge Discovery*, 1(5):397–415, 2011.
13. Y. Feng, M. Ji, J. Xiao, X. Yang, J. J. Zhang, Y. Zhuang, and X. Li. Mining spatial-temporal patterns and structural sparsity for human motion data denoising. *IEEE Transactions on Cybernetics*, 45(12):2693–2706, 2015.
14. A. A. Freitas. *Data Mining and Knowledge Discovery with Evolutionary Algorithms*. Springer-Verlag Berlin Heidelberg, 2002.
15. H. Gao, S. Shiji, J.N.D. Gupta, and W. Cheng. Semi-supervised and unsupervised extreme learning machines. *IEEE Transactions on Cybernetics*, 44(12):2405–2417, 2014.
16. M. Gendreau and J. Potvin. *Handbook of Metaheuristics*. Springer Publishing Company, Incorporated, 2nd edition, 2010.
17. T. George, M. Nathan, M. Wagner, and F. Renato. Tree projection-based frequent itemset mining on multi-core CPUs and GPUs. In *Proceedings of the 22nd International Symposium on Computer Architecture and High Performance Computing*, SBAC-PAD 2010, pages 47–54, Petrópolis, Brazil, October 2010.
18. S. Ghemawat, H. Gobioff, and S. Leung. The google file system. In *Proceedings of the Nineteenth ACM Symposium on Operating Systems Principles*, SOSP '03, pages 29–43, New York, NY, USA, 2003. ACM.
19. B. Goethals and M.J. Zaki. Advances in frequent itemset mining implementations: report on fimi'03. *ACM SIGKDD Explorations Newsletter*, 6(1):109–117, 2004.
20. B. Goethals, S. Moens, and J. Vreeken. MIME: A Framework for Interactive Visual Pattern Mining. In D. Gunopulos, T. Hofmann, D. Malerba, and M. Vazirgiannis, editors, *Machine Learning and Knowledge Discovery in Databases*, volume 6913 of *Lecture Notes in Computer Science*, pages 634–637. Springer Berlin Heidelberg, 2011.
21. R. C. Green II, L. Wang, M. Alam, and R. A. Formato. Central force optimization on a GPU: A case study in high performance metaheuristics. *Journal of Supercomputing*, 62(1):378–398, 2012.
22. J. Han and M. Kamber. *Data Mining: Concepts and Techniques*. Morgan Kaufmann, 2000.
23. J. Han, J. Pei, Y. Yin, and R. Mao. Mining Frequent Patterns without Candidate Generation: A Frequent-Pattern Tree Approach. *Data Mining and Knowledge Discovery*, 8:53–87, 2004.
24. H. Kwasnicka and K. Switalski. Discovery of association rules from medical data: classical and evolutionary approaches. *Annales UMCS, Informatica*, 4(1):204–217, 2006.
25. W. B. Langdon. Performing with CUDA. In *Proceedings of the 13th annual Genetic and Evolutionary Computation Conference*, GECCO 2011, pages 423–430, Dublin, Ireland, 2011.
26. R. W. P. Luk and W. Lam. Efficient in-memory extensible inverted file. *Information Systems*, 32(5):733–754, 2007.
27. J. M. Luna, J. R. Romero, and S. Ventura. G3PARM: A Grammar Guided Genetic Programming Algorithm for Mining Association Rules. In *Proceedings of the IEEE Congress on Evolutionary Computation*, IEEE CEC 2010, pages 2586–2593, Barcelona, Spain, 2010.
28. J. M. Luna, J. R. Romero, and S. Ventura. Design and behavior study of a grammar-guided genetic programming algorithm for mining association rules. *Knowledge and Information Systems*, 32(1):53–76, 2012.
29. J. M. Luna, J. R. Romero, C. Romero, and S. Ventura. Reducing gaps in quantitative association rules: a genetic programming free-parameter algorithm. *Integrated Computer Aided Engineering*, 21(4):321–337, 2014.
30. J. M. Luna, A. Cano, M. Pechenizkiy, and S. Ventura. Speeding-Up Association Rule Mining With Inverted Index Compression. *IEEE Transactions on Cybernetics*, pp(99):1–14, 2016.
31. D. Martín, A. Rosete, J. Alcalá, and F. Herrera. A new multiobjective evolutionary algorithm for mining a reduced set of interesting positive and negative quantitative association rules. *IEEE Transactions on Evolutionary Computation*, 18(1):54–69, 2014.

32. D. Martín, A. Rosete, J. Alcalá-Fdez, and F. Herrera. Qar-cip-nsga-ii: A new multi-objective evolutionary algorithm to mine quantitative association rules. *Information Sciences*, 258:1–28, 2014.

33. D. Martín, M. Martínez-Ballesteros, S. Río, J. Alcalá-Fdez, J. Riquelme, and F. Herrera. MOPNAR-BigData: un diseno MapReduce para la extracción de reglas de asociación cuantitativas en problemas de Big Data. In *Actas de la XVI Conferencia de la Asociación Española para la Inteligencia Artificial*, CAEPIA 2015, pages 979–989, Albacete, Spain, November 2015.

34. M. Martinez-Ballesteros, S. Salcedo-Sanz, J. C. Riquelme, C. Casanova-Mateo, and J. L. Camacho. Evolutionary association rules for total ozone content modeling from satellite observations. *Chemometrics and Intelligent Laboratory Systems*, 109(2):217–227, 2011.

35. V. Marx. The big challenges of big data. *Nature*, 498(7453):255–260, 2013.

36. J. Mata, J. L. Alvarez, and J. C. Riquelme. Mining numeric association rules with genetic algorithms. In *Proceedings of the 5th International Conference on Artificial Neural Networks and Genetic Algorithms*, ICANNGA 2001, pages 264–267, Taipei, Taiwan, 2001.

37. J. Mata, J. L. Alvarez, and J. C. Riquelme. Discovering numeric association rules via evolutionary algorithm. In *Proceedings of the 6th Pacific-Asia Conference on Advances in Knowledge Discovery and Data Mining*, PAKDD 2002, pages 40–51, Taipei, Taiwan, 2002.

38. R. McKay, N. Hoai, P. Whigham, Y. Shan, and M. O'Neill. Grammar-based Genetic Programming: a Survey. *Genetic Programming and Evolvable Machines*, 11:365–396, 2010.

39. S. Moens, E. Aksehirli, and B. Goethals. Frequent itemset mining for big data. In *Proceedings of the 2013 IEEE International Conference on Big Data*, pages 111–118, Santa Clara, CA, USA, October 2013.

40. J. L. Olmo, J. M. Luna, J. R. Romero, and S. Ventura. Mining association rules with single and multi-objective grammar guided ant programming. *Integrated Computer-Aided Engineering*, 20(3):217–234, 2013.

41. F. Pulgar-Rubio, C. J. Carmona, A. J. Rivera-Rivas, P. González, and M. J. del Jesus. MUna primera aproximación al descubrimiento de subgrupos bajo el paradigma MapReduce. In *Actas de la XVI Conferencia de la Asociación Española para la Inteligencia Artificial*, CAEPIA 2015, pages 991–1000, Albacete, Spain, November 2015.

42. Y. Qian, J. Liang, W. Pedrycz, and C. Dang. An efficient accelerator for attribute reduction from incomplete data in rough set framework. *Pattern Recognition*, 44(8):1658–1670, 2011.

43. D. Wegener, M. Mock, D. Adranale, and S. Wrobel. Toolkit-based high-performance data mining of large data on MapReduce clusters. In *Proceedings of the IEEE International Conference on Data Mining*, ICDM 2009, pages 296–301, Miami, Florida, USA, 2009.

44. F. Wenbin, L. Mian, X. Xiangye, H. Bingsheng, and L. Qiong. Frequent itemset mining on graphics processors. In *Proceedings of the 5th International Workshop on Data Management on New Hardware*, DaMoN '09, pages 34–42, Providence, Rhode Island, 2009.

45. X. Yan, C. Zhang, and S. Zhang. Genetic algorithm-based strategy for identifying association rules without specifying actual minimum support. *Expert Systems with Appications*, 36:3066–3076, 2009.

46. M. J. Zaki. Scalable algorithms for association mining. *IEEE Transactions on Knowledge and Data Engineering*, 12(3):372–390, 2000.

47. C. Zhang and S. Zhang. *Association rule mining: models and algorithms*. Springer Berlin / Heidelberg, 2002.

48. J. Zhou, K. M. Yu, and B. C. Wu. Parallel frequent patters mining algorithm on GPU. In *Proceedings of the IEEE International Conference on Systems, Man and Cybernetics*, SMC 2010, pages 435–440, Istanbul, Turkey, 2010.

Printed in the United States
By Bookmasters